SECOND EDITION

NATURAL RESOURCE MEASUREMENTS AND MAPPING

An Introduction to Field Skills and Data Analysis

ALEXANDRA LOCHER
GRAND VALLEY STATE UNIVERSITY

cognella
SAN DIEGO

Bassim Hamadeh, CEO and Publisher
Carrie Montoya, Manager, Revisions and Author Care
Kaela Martin, Project Editor
Alia Bales, Production Editor
Jess Estrella, Senior Graphic Designer
Alexa Lucido, Licensing Manager
Natalie Piccotti, Director of Marketing
Kassie Graves, Vice President of Editorial
Jamie Giganti, Director of Academic Publishing

Printed in the United States of America.

cognella® | ACADEMIC PUBLISHING
3970 Sorrento Valley Blvd., Ste. 500, San Diego, CA 92121

Dedicated to my students

"Skill'd in the globe and sphere,
he gravely stands,
And, with his compass,
measures seas and lands."

—JOHN DRYDEN

Contents

Foreword

Wouldn't it be awesome to hop on a charter boat and fish for salmon, or hike through a forest while tracking a radio-collared animal to document habitat use? Or perhaps it would be amazing to search for endangered plants and create maps of conservation areas, inventory trees, collect water samples in a river, or band migratory birds. What if you got *paid* for doing those things? The good news is that with increasing global awareness and desires to conserve our earth, natural resources–based fields of study are becoming increasingly important. Thus, job opportunities working outdoors or inside analyzing environmental data are available for qualified candidates. On recent job boards advertising natural resources–related jobs, skills in field measurements, geographic information systems, data organization and analysis, global positioning systems, map reading and navigation, remote sensing, and problem solving appear as qualifications in many of the posted entry-level positions. But, those jobs are competitive! Thus, those seeking employment in natural resources–related jobs should be proficient in such skills.

This book was designed to provide a foundation for many of the important technical and analytical skills required by many natural resources professions. For instance, compass use and navigation is essential for finding your way during field activities. Because natural resources–related fields investigate the earth and human interactions with the earth, understanding how to interpret maps and take horizontal and vertical measurements from them is imperative for success. Many jobs in this field require an assessment of property boundaries or topographic characteristics of an area. Therefore, understanding legal descriptions from boundary surveys and conducting topographic surveys are also important skills. Skills processing earth data such as imagery collected from satellites, using global positioning systems to record positions of important features, or using computer software programs to analyze spatial relationships between features on the earth are becoming increasingly marketable.

This book is intended to serve as a fundamental textbook and workbook to introduce you to many practical skills associated with natural resources–related fields. It is my sincere intention and hope that this book may inspire and provide some foundational skills to launch and help direct your natural resources education. There is so much to learn! Have fun on your journey!

Ali Locher
January 2016

1

Compass, Maps, and Location

In many fields and disciplines, understanding maps, compass use, and methods of determining position is critical. Many jobs—such as those working for state and federal agencies related to fisheries, wildlife, forestry, geology, or natural resources—require employees to be skilled in map reading, making measurements from maps, and determining position. In this chapter, you will learn an introduction to tools involved in determining location. **Location** refers to a particular place or position.

LEARNING OBJECTIVES

1. Define *compass*
2. Differentiate between latitude and longitude; explain their role in determining location
3. Explain the purpose of maps
4. Compute location using sun angle, time of day, Greenwich Mean Time, and analemma correction

Compass and North

A **compass** is an instrument used for navigation. It shows the four cardinal directions (north, east, south, and west) relative to the surface of the earth. A magnetic needle on the compass points to magnetic north. The magnetic compass was invented in China over 2,000 years ago. Early compasses used lodestone, a lump of magnetic rock, to magnetize iron needles. Once magnetized, the needles would orient in a north–south direction.

Compass needles point in a northern direction because the earth's magnetic field is polarized, which means that positive and negative charges are separated. This polarization occurs because the outer core of the earth is composed of liquid metals, including iron and nickel. As the earth spins, the molten metal flows and creates an electric current. That current creates the magnetic field surrounding the planet. Thus, a compass needle is attracted to the magnetic charge in the northern part of the earth (i.e., magnetic north).

One of the first things to understand about map reading, location, and compass use is the difference between three "norths": magnetic north, true north, and grid north. **Magnetic north** is a location in the northern polar region that has a strong electromagnetic charge. This location is currently more than 700 miles from true north. **True north** is the location where lines of longitude converge. The difference between magnetic north and true north is **declination**. When using maps and a compass in the field, it is imperative to consider declination for accurate navigation. We will cover declination in detail in Chapter 4. **Grid north** refers to the north of a particular geographic area represented on a map. It approximates true north, but its accuracy depends on particular map projections.

Maps

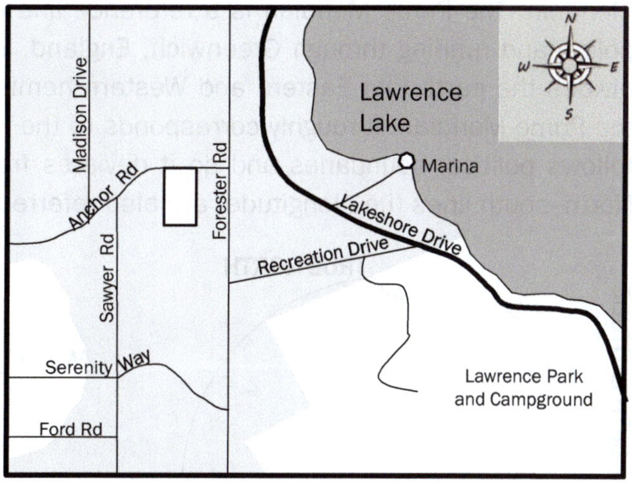

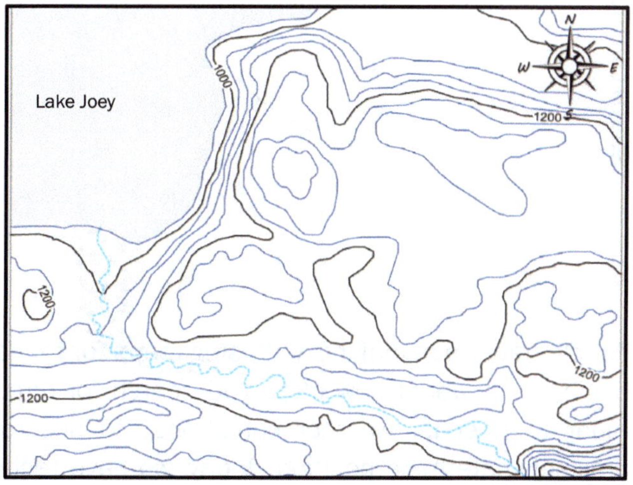

Maps are models of some portion of the earth's surface. Maps generalize features and are intended to depict specific phenomena within a geographic area. Maps are used for many purposes. First, they may represent large areas. For instance, they may depict features within the world's continents or entire countries. Maps can show information we cannot readily see. For example, maps may depict temperatures, average precipitation, population densities, wildlife habitat suitability, or other modeled information. Maps show spatial relationships and present information concisely. From any map, one is able to understand spatial relationships: is the river flowing east or west? Which roads take me from A to B?

There are many types of maps, but generally they can fit into two categories: planimetric or topographic. **Planimetric maps** illustrate only the horizontal positions of features with no reference to contours, elevation, or other topographic relief. Alternatively, **topographic maps** represent hills, valleys, and other features of the earth's terrain.

Latitude and Longitude

Latitude and longitude are angular measurements used for determining location on the surface of the earth. **Latitude** is the angular distance measured in degrees north or south of the equator. Latitude is used to refer to a position in the northern or southern hemisphere. The equator is 0° latitude, and divides the earth into Northern and Southern hemispheres. The North Pole is 90° North latitude. The South Pole is 90° South latitude (or −90°). Lines of latitude are also referred to as parallels. For instance, if you were standing on the 45th parallel, you would be located halfway between the equator and one of the poles. **Longitude** is the angular distance measured in degrees east or west of the Prime Meridian. The Prime Meridian is a reference line extending between the north and south poles, and running through Greenwich, England. The Prime Meridian is 0° longitude and divides the earth into Eastern and Western hemispheres. The Antimeridian is 180° from the Prime Meridian. It roughly corresponds to the International Date Line (IDL), but the IDL follows political boundaries and so it deviates from the Antimeridian in some locations. North–south lines (i.e., longitude) are also referred to as meridians.

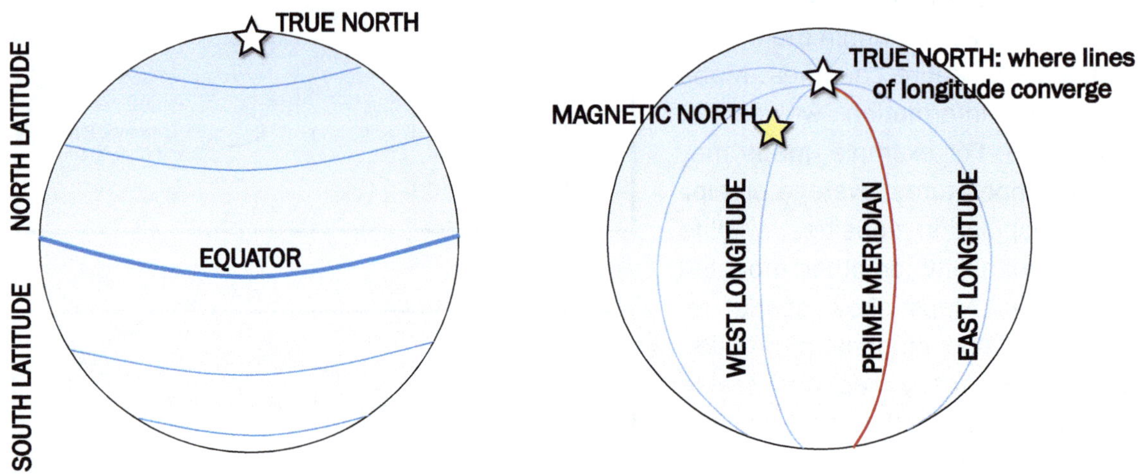

Methods to Determine Location

People have been making and using maps for thousands of years. Maps are used on a daily basis. Think about how often you search for directions to a location or view the Doppler radar to forecast weather conditions. To accurately create maps, cartographers must precisely determine position. Today, we have technological advancements such as Global Positioning Systems that automatically determine position. Prior to modern technology, cartographers used points of reference—such as the position of the sun or the stars—to determine latitude. Longitude was determined from local time in relation to known time at a specific meridian, such as the Prime Meridian. The purpose of this section is to help you think of position on the earth in terms of latitude and longitude. This basic knowledge will help prepare you for using maps, navigating with a compass, and understanding Global Positioning Systems and Geographic Information Systems in later chapters.

Latitude

To determine latitude, you need to know the date, the sun's angle at midday (i.e., local noon), and the declination of the sun (i.e., the angle of the sun in relation to the equator). An analemma is a figure-eight pattern that the sun makes if you were to observe it at the same time from the same location throughout the year. Because the earth is tilted on its axis and the earth's rotation around the sun is not a perfect circle, the sun's position appears to take a figure-eight pattern (an **analemma**) annually. Graphically representing the position of the sun throughout the year in an analemma can help people understand the angle of the sun in relation to the equator and the equation of time, which is the difference between mean solar time (the time on your clock) and apparent solar time (time measured from a sundial). During some parts of the year, the sun is apparently faster than what our clocks indicate. During other parts of the year, the sun is apparently slower than what our clocks indicate.

The following equation is for the calculation of latitude:

$$\text{Latitude} = (90° - \text{solar angle at midday}) \pm \text{declination of the sun}$$

Where,

> Solar angle at midday = the angle of the sun at its highest point at your location. This angle is measured from the ground to the sun. A **sextant** was historically used to measure the altitude of the sun above the horizon. If the sun was directly overhead, the angle was 90°. If the sun was at the horizon, the angle was 0°.
>
> Declination of the sun = the angle of the sun in relation to the equator, as indicated on an analemma (see figure on the next page). The sun's declination is 0° when it is directly above the equator (i.e., during the spring and fall solstices). If you are in the northern hemisphere and the sun is north of the equator, add the declination. If the sun is south of the equator, subtract declination.

For example, pretend you are in the northern hemisphere on October 20. At your location, the sun is at an angle of 38° at its highest point in the sky. So, we know that the solar angle at midday is 38°.

According to the analemma, on October 20, the sun's declination is 9° south of the equator, or −9°. Using the equation, we calculate latitude as

$$\text{Latitude} = (90° - 38°) - 9°$$

$$\text{Latitude} = 43° \text{ North}$$

Latitude is positive in the northern hemisphere, and negative in the southern hemisphere.

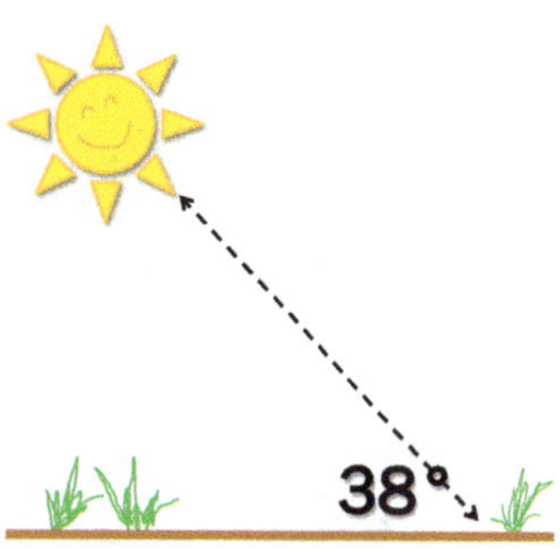

Date: October 20

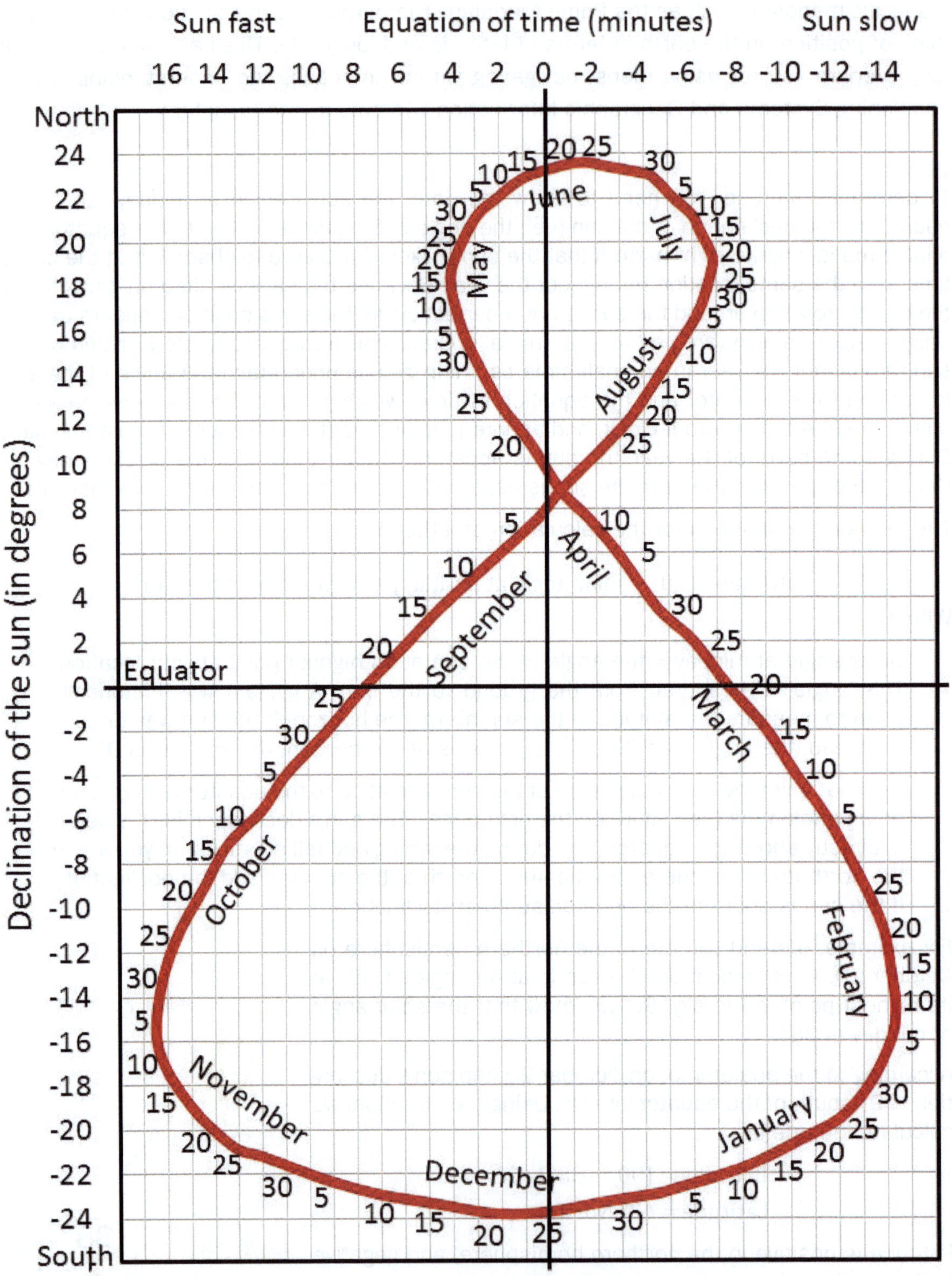

Source: Institut de Physique du Globe de Paris. Data Source: U.S. Coast and Geodetic Survey.

Longitude

To determine longitude, you need to know the relationship between time and the rotation of the earth. The sun passes east to west over a given meridian every 24 hours. In relation to a given meridian, local time EAST of the meridian is later; local time WEST of the meridian is earlier. For example, if you were standing on the Prime Meridian at noon, it would be afternoon (pm) in locations to the east of you. Likewise, it would be morning (am) in locations to the west of you.

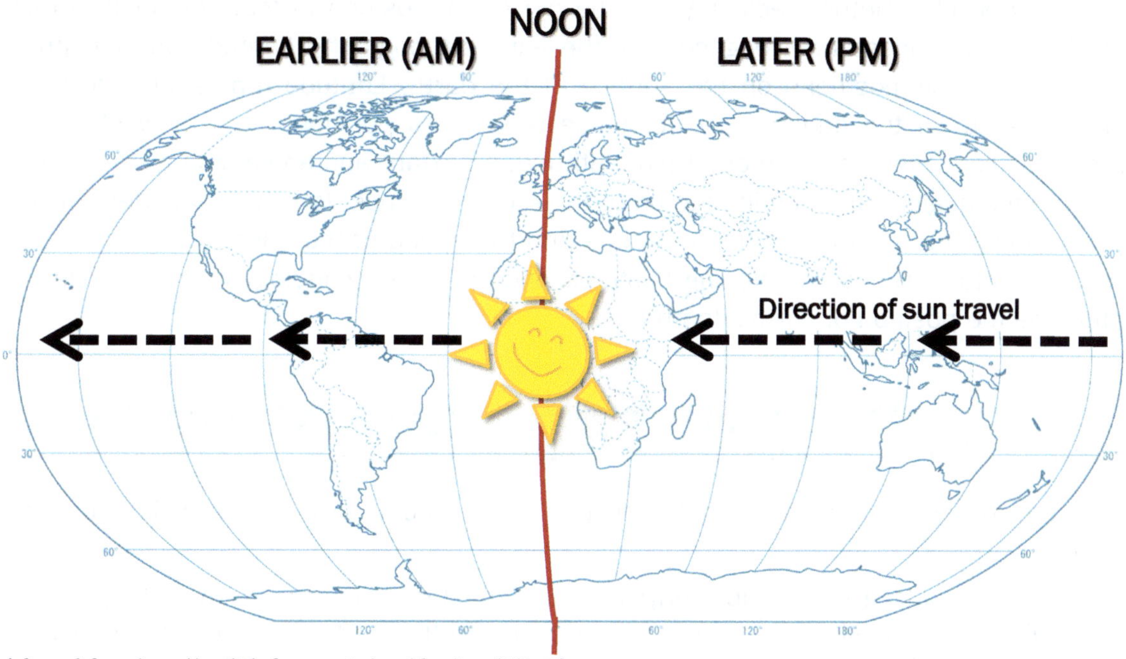

Adapted from http://english.freemap.jp/world_e/img/4/2.gif

So, because the sun passes over 360° every 24 hours, each HOUR it passes over 15° of longitude (360°/24 hours). Therefore, each MINUTE it passes over 0.25° longitude. If you knew the time at a specific meridian, you could calculate your position based on the time difference between your location and the known time at the specific meridian. A **chronometer** is another navigation instrument invented to keep extremely accurate time at known locations. For example, **Greenwich Mean Time (GMT)** is the mean solar time on the Prime Meridian (i.e., 0° longitude). It was commonly used as a time standard and in calculations of position relative to the Prime Meridian.

Suppose you were lost and wanted to determine your position. Suppose also that you knew GMT. If it was noon at your location (when the sun was at its highest position in the sky), and it was 3:00 pm GMT, at what longitude are you?

Well, in this case it is earlier at your location (noon) than GMT (3:00 pm). So, you should know that you are located *west* of the Prime Meridian. Your longitude should be negative. The difference in time between you and GMT is 3 hours. Thus, the sun has traversed 45° longitude (3 hours * 15° longitude/hour). Therefore, you know that you are located at 0° – 45° = –45°, or 45° West of the Prime Meridian.

SUMMARY

A compass is an important navigational instrument that shows the four cardinal directions. Maps may be planimetric, depicting only the horizontal position of features, or they may depict the smoothness and roughness of the earth's topography. Latitude and longitude are used to show positions on the surface of the earth. Latitude represents positions north or south of the equator, and longitude represents positions east or west of the Prime Meridian. Historically, to determine latitude, a sextant and analemma were used to measure the angle of the sun in relation to the earth's surface; longitude was determined by calculating the relationship between time and the speed of the earth's rotation. Today, advancements in Global Positioning Systems and satellite technology make it easier to determine the position of earth's features.

TEST YOUR SKILLS

Solve the following problems to help you think about position in terms of latitude and longitude.

1. You are working on a ship studying beluga whales. You have been sailing around collecting population data for conservation efforts. In one location, you located a particularly large pod of whales and want to mark the location. For some reason, your GPS is not receiving a satellite signal, and the ship's navigator is sick in bed with a terrible virus. Basically, no one really knows the current position of the ship. So, you resort to your knowledge on positioning. It's June 10, and at local noon, the angle of the sun is only 38°. Your watch, which is set to Greenwich Mean Time, indicates that it is 10:00 pm when it is noon where you are. What are the latitude–longitude coordinates of the beluga whale pod?

2. You are working in Canada studying tundra wildlife. The most efficient form of travel between study sites is by fixed-wing aircraft. While traveling from one study site due north to another site, a freak electrical current struck the plane and wiped out its navigation system. The pilot landed, but you're not really sure where you are located. So, you resort to your knowledge on positioning. It's July 18, and at local noon, the angle of the sun is 53°. Before the navigation system broke, you noticed that your local time was 7.5 hours earlier than Greenwich Mean Time. What are your latitude–longitude coordinates?

3. You were on vacation in London, England. You hired a pilot with a small fixed-wing aircraft to fly you to Paris, France. An unexpected storm blew you off course and you landed safely, but the pilot is not sure where you are. It is April 12 and the sun's angle at midday where you are is 42.5°. When the sun is at the highest point in the sky at your location, your watch, which is set to GMT, reads 11:10 am. What is your latitude and longitude? Where are you?

REFERENCES

Campbell, J. (2001). *Map Use & Analysis*, 4th Edition. New York: NY McGraw-Hill Higher Education.

Jacobson, C. (1999). *Basic Essentials: Map and Compass*, 2nd Edition. Guilford, CT: The Globe Pequot Press.

Singh, R., Artman, D., Tayor, D. W., & Brinton, D. (2000). *Basic Surveying: Theory and Practice*. Ninth Annual Seminar Presented by the Oregon Department of Transportation Geometronics Unit. Bend, OR. http://www.oregon.gov/odot/hwy/geometronics/docs/basicmanual2000_02.pdf

"A Short Guide to Celestial Navigation." http://www.celnav.de/

2

Scale and Conversions

Map use requires a clear understanding of scale. **Scale** is the ratio between map distance and earth distance. Often map scales are represented as ratios (without units) or in distance units such as feet, miles, or kilometers that may need to be converted to other units. Occasionally, map users may need to determine area of features on maps. Proficiency in map measurements using scale and conversions is critical for developing your skills as a natural resources professional.

LEARNING OBJECTIVES

1. Define *scale*
2. Explain the difference between a representative fraction and an equivalent scale
3. Convert between representative fractions and equivalent scales
4. Recognize common English and metric units of linear measurements
5. Recognize common English and metric units of area measurements
6. Convert between various units of linear measurements
7. Convert between various units of area measurements

Scale

There are two general methods of expressing scale. An **equivalent scale** uses two units to describe the relationship between map distance and earth distance. For example, an equivalent scale might indicate that 1 inch on the map equals 4 miles on the ground. Equivalent scales may be depicted in words (e.g., 1 inch to 4 miles) or graphics. Graphic scales are lines drawn on maps and subdivided into units appropriate to the scale.

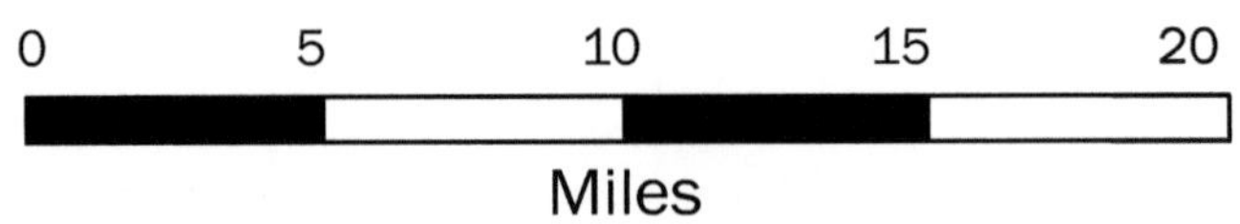

A **representative fraction** represents scale as the ratio of distance on the map to the same distance on the ground. For instance, a scale of 1:24,000

means that 1 unit on the map is equivalent to 24,000 of the same units on the ground. A ratio does not have units. It is a relationship that holds true regardless of what units are used. So, 1 centimeter on the map represents 24,000 centimeters on the surface of the earth. By the same token, 1 inch on the same map represents 24,000 inches on the surface of the earth.

You may have heard the terms **small scale** and **large scale** as they refer to maps. These terms refer to the ratio between map and earth, or the representative fraction. A small-scale map has a very small ratio. For instance, a scale of 1:24,000 or smaller would be considered small scale. If you divide 1/24,000, you would get a very small number (0.0000417). Alternatively, a large-scale map has a larger ratio, such as 1:1,000 or larger. Dividing 1/1,000 yields 0.001, a much larger number than 0.0000417.

Often, the terms "small scale" and "large scale" are used incorrectly. Be aware of this. Sometimes you will hear people refer to a large-scale map as a map that represents a large spatial extent. Not correct! Likewise, you will hear people refer to a small-scale map as a map that represents a small spatial extent. Again, not correct! An easy way to differentiate between small- and large-scale maps is to look at the features on the maps. Features or objects on small-scale maps are small; the ratio between map distance and earth distance is quite small, so small-scale maps represent large spatial extents. Features or objects on large-scale maps are larger, so they represent smaller spatial extents.

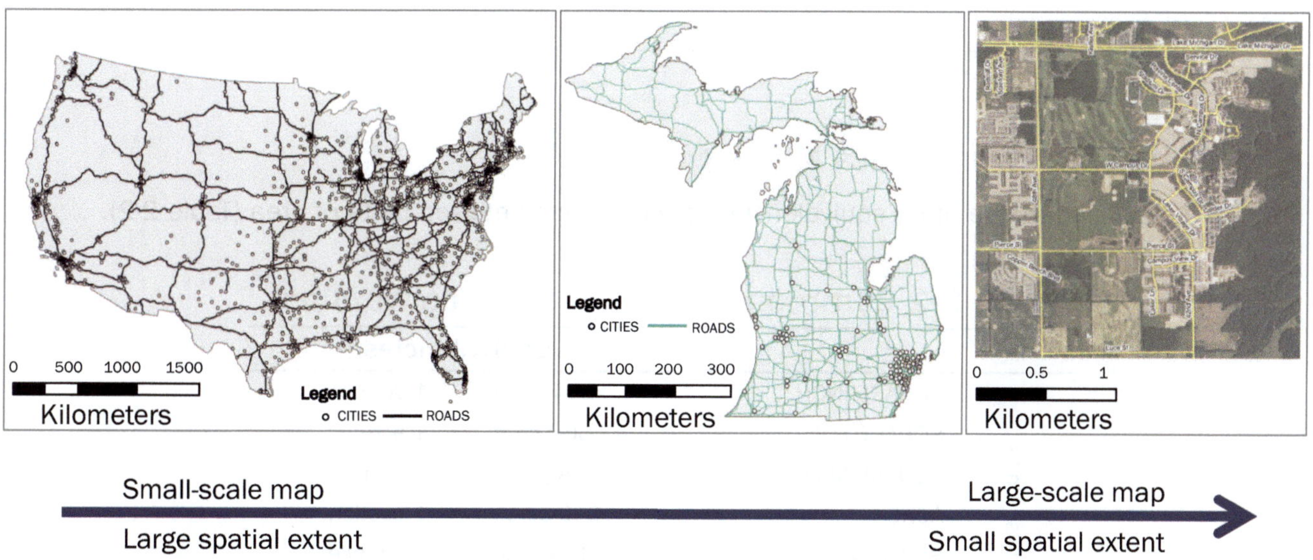

(a) Tigerline Primary and Secondary Roads, U.S. Census Bureau.
(b) NRCS States by State, U.S. Department of Agriculture.
(c) 2014 National Ag. Imagery Program Mosaic, U.S. Department of Agriculture.

Conversions: Common Linear Measurements

Although you can typically look up conversions, it is helpful to learn them so you can compute conversions quickly and confidently (Table 2-1). You may need to use these equivalencies for converting between units of scale.

Table 2-1 Useful and common linear conversions

	Unit	Equivalencies		
English Units	INCH	12 in	=	1 ft
	FOOT	3 ft	=	1 yd
	YARD	1 yd	=	3 ft
	ROD	1 rod	=	16.5 ft
	CHAIN	1 chain	=	66 ft
	MILE	1 mi	=	5,280 ft
		1 mi	=	80 chains
Metric Units	MILLIMETER	1 mm	=	0.001 m
	CENTIMETER	1 cm	=	0.01 m
	DECIMETER	1 dm	=	0.1 m
	METER			
	DEKAMETER	1 dkm	=	10 m
	HECTOMETER	1 hm	=	100 m
	KILOMETER	1 km	=	1,000 m

Conversions: Common Area Measurements

You may need to use other equivalencies for converting between units of area (Table 2-2).

Table 2-2 Common and useful area conversions

		Equivalencies		
English Units	SQUARE INCH	144 sq in	=	1 sq foot
	SQUARE FOOT	43,560 sq ft	=	1 acre
	SQUARE MILE	1 sq mile	=	640 acres
	ACRE	640 acres	=	1 section
	SECTION	1 section	=	1 sq mile
	TOWNSHIP	1 twp	=	36 sections
		1 twp	=	36 sq miles

Metric Units	SQUARE CENTIMETER	1 sq cm	=	0.0001 sq m
	SQUARE METER			
	ARE	1 are	=	100 sq m
	DECARE	1 da	=	1,000 sq m
	HECTARE	1 ha	=	10,000 sq m

Conversions: Common Conversions between English and Metric Units

The following conversions are very useful to learn because it's likely you will encounter them frequently in your profession.

Table 2-3 Useful and common conversions between English and metric units

1	MILE	=	1.61	KILOMETERS
1	KILOMETER	=	0.621	MILES
1	ACRE	=	0.405	HECTARES
1	HECTARE	=	2.47	ACRES
1	INCH	=	2.54	CENTIMETERS
1	CENTIMETER	=	0.394	INCH
1	METERS	=	3.281	FEET

Scale Conversions

Converting between an equivalent scale and a representative fraction is relatively straightforward, but involves your knowledge of algebra. It is important to keep track of units when converting. Let's solve the following problem:

► Convert an equivalent scale of 1 inch = 2 miles to a representative fraction.

1. You know that a representative fraction is irrespective of units. Therefore, both parts of the fraction need to be in the same units so that they cancel. The scale can be written as follows:

$$\frac{1\ inch}{2\ miles}$$

2. Next, to get both the numerator and the denominator into the same units, it makes sense to convert the denominator into inches, so the units cancel. There are 5,280 feet per mile and 12 inches per foot (or 63,360 inches per mile).

$$\frac{1\ \cancel{inch}}{2\ \cancel{miles}} * \frac{1\ \cancel{mi}}{5{,}280\ \cancel{ft}} * \frac{1\ \cancel{ft}}{12\ \cancel{in}} = \frac{1}{126{,}720}$$

3. Note that all the units in the numerator cancel with units in the denominator, so you are left with a ratio, or representative fraction. The representative fraction 1:126,720 is equivalent to 1 inch = 2 miles.

In cases when the numerator is not equal to 1, you may need to perform an additional step to calculate a representative fraction from a given equivalent scale. For example:

▶ Find a representative fraction for an equivalent scale of 6 inches = 1 mile.

1. First, determine the miles equivalent to 1 inch by dividing the numerator and denominator by 6:

$$\frac{6\ inches}{1\ mile} = \frac{1\ inch}{0.167\ miles}$$

2. Next, to get both the numerator and the denominator into the same unit, convert the denominator into inches, so the units cancel.

$$\frac{1\ \cancel{inch}}{0.167\ \cancel{miles}} * \frac{1\ \cancel{mi}}{5{,}280\ \cancel{ft}} * \frac{1\ \cancel{ft}}{12\ \cancel{in}} = \frac{1}{10{,}560}$$

3. Write your answer as a ratio. The representative fraction equivalent to a scale of 6 inches = 1 mile is 1:10,560.

Now, let's practice converting a representative fraction to an equivalent scale.

▶ A map indicates a scale of 1:55,000. According to this scale, how many kilometers on the earth are equivalent to 1 centimeter on the map?

1. Recall that a representative fraction is a ratio of map distance to earth distance in the same units. Therefore, you can assign the numerator and denominator any units, as long as they are the same. Thus, our scale can be interpreted as 1 centimeter on the map is equivalent to 55,000 centimeters on the earth.

$$1\ cm = 55{,}000\ cm$$

2. Now convert 55,000 centimeters to kilometers. Pay attention to the units and make sure the appropriate ones cancel. There are 100 centimeters per meter and 1,000 meters per kilometer.

$$\frac{55{,}000\ \cancel{cm}}{1} * \frac{1\ \cancel{m}}{100\ \cancel{cm}} * \frac{1\ km}{1{,}000\ \cancel{m}} = 0.55\ km$$

3. A representative fraction of 1:55,000 equals an equivalent scale of 1 cm = 0.55 km.

Determining Area Using Scale

Suppose you needed to determine area of a land parcel on a map. If you can measure the dimensions of the land parcel, and you know the scale, then you can determine area. For example, if a parcel of land on a map is 1 inch x 1 inch and the map scale is 1:6,000, how many acres are in the land parcel?

1. It's always helpful to draw a diagram of the problem. Visualizing the problem gets you one step closer to solving it. The parcel measures one square inch on the map. One inch on the map is equivalent to 600 inches on the ground.
2. Ultimately we want to determine the number of acres of the parcel. So a logical next step is to convert 6,000 inches to feet by dividing by 12 (12 inches per foot); 6,000 inches is equivalent to 500 ft.
3. Now, you are ready to solve for area. The area of a square or rectangle equals length times width.

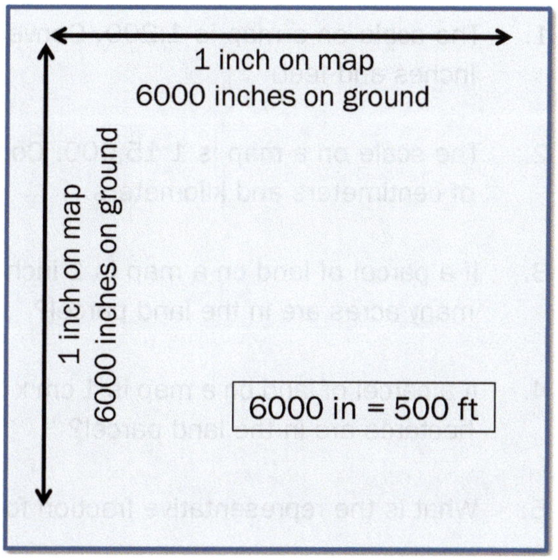

$$500\ ft * 500\ ft = 250{,}000\ ft^2$$

4. Finally, you can convert the area in square feet to area in units of acres. There are 43,560 ft^2 per acre (Table 2). So, the area of this land parcel is approximately 5.7 acres.

$$\frac{250{,}000\ ft^2}{1} * \frac{1\ ac}{43{,}560\ ft^2} = 5.7\ ac$$

SUMMARY

Map use requires a clear understanding of scale, which is the ratio between the distance measurements on maps and actual distance measurements on the surface of the earth. Scale may be depicted on maps as equivalent scales or representative fractions. On small scale maps, features on the map are small (i.e., as if you were "zoomed out") because the ratio between map and earth distance is small. Large scale maps have relatively larger features (i.e., as if you were "zoomed in") because the ratio between map distance and earth distance is larger. Scale is important to consider when making accurate linear and area measurements. Often, you may need to convert different linear (e.g., centimeters, inches, feet, kilometers) or area (e.g., square inches, square meters, acres, hectares) units in order to interpret map relationships.

TEST YOUR SKILLS

1. The scale on a map is 1:200. Convert this scale to an equivalent scale in terms of inches and feet.

2. The scale on a map is 1:15,000. Convert this scale to an equivalent scale in terms of centimeters and kilometers.

3. If a parcel of land on a map is 1 inch x 2 inches and the map scale is 1:1,000, how many acres are in the land parcel?

4. If a parcel of land on a map is 1 cm x 1 cm and the map scale is 1:50,000, how many hectares are in the land parcel?

5. What is the representative fraction for an equivalent scale of 1 inch = 0.5 mile?

6. What is the representative fraction for an equivalent scale of 1 cm = 1 kilometer?

7. Convert the following:

   a. 12 km = ______________ m

   b. 11 mi = ______________ m

   c. 5 mi = ______________ yds

d.	16 in	=	________________	mm
e.	2.54 cm	=	________________	m
f.	5 km	=	________________	ft
g.	2.8 ft	=	________________	in
h.	6,000 ft	=	________________	km
i.	120 mm	=	________________	cm
j.	4.5 ac	=	________________	ha
k.	1 mi^2	=	________________	ac
l.	2.5 mi^2	=	________________	ha

REFERENCE

Campbell, J. (2001). *Map Use & Analysis*, 4th Edition. New York, NY: McGraw-Hill Higher Education.

3

Pacing and Basic Compass Use

Orienteering is a form of navigation requiring skills in the use of maps and compasses. **Navigation** is a skill that involves determination of position and direction. Understanding how to use a map and compass together to navigate is a critical skill for natural resources professionals, as maps and field skills are used widely in this profession. For example, during your career, you may need to navigate to specific locations plotted on a map to sample vegetation, collect water samples, or study a wildlife species.

In this exercise, you will understand the basics of using maps and a compass for navigation. It will provide the foundation for developing your spatial skills.

LEARNING OBJECTIVES

1. Learn the anatomy of a compass
2. Understand the difference between a bearing and an azimuth and convert between them
3. Know your pace and how to use it to find distances on the ground
4. Understand how to acquire an azimuth from your position to an object in the field
5. Understand how to orient your compass, given a specific azimuth

Anatomy of a Compass

There are several different models of compasses, but the basic anatomy is similar among all of them. The **Scale** or **Ruler** allows you to measure distances from maps. The **Direction of Travel Arrow** should always point to your destination. The **Baseplate** is the structure on which the compass is mounted. Direction of travel in the form of an azimuth or bearing should always be read from the **Index Line**. The **Magnetic Needle** points to magnetic north in the field. The **Orienting Lines** indicate true north–south lines on a map. The **Degree Dial** houses the compass and can rotate to align the **Orienting Arrow** with the Magnetic Needle to find your direction of travel.

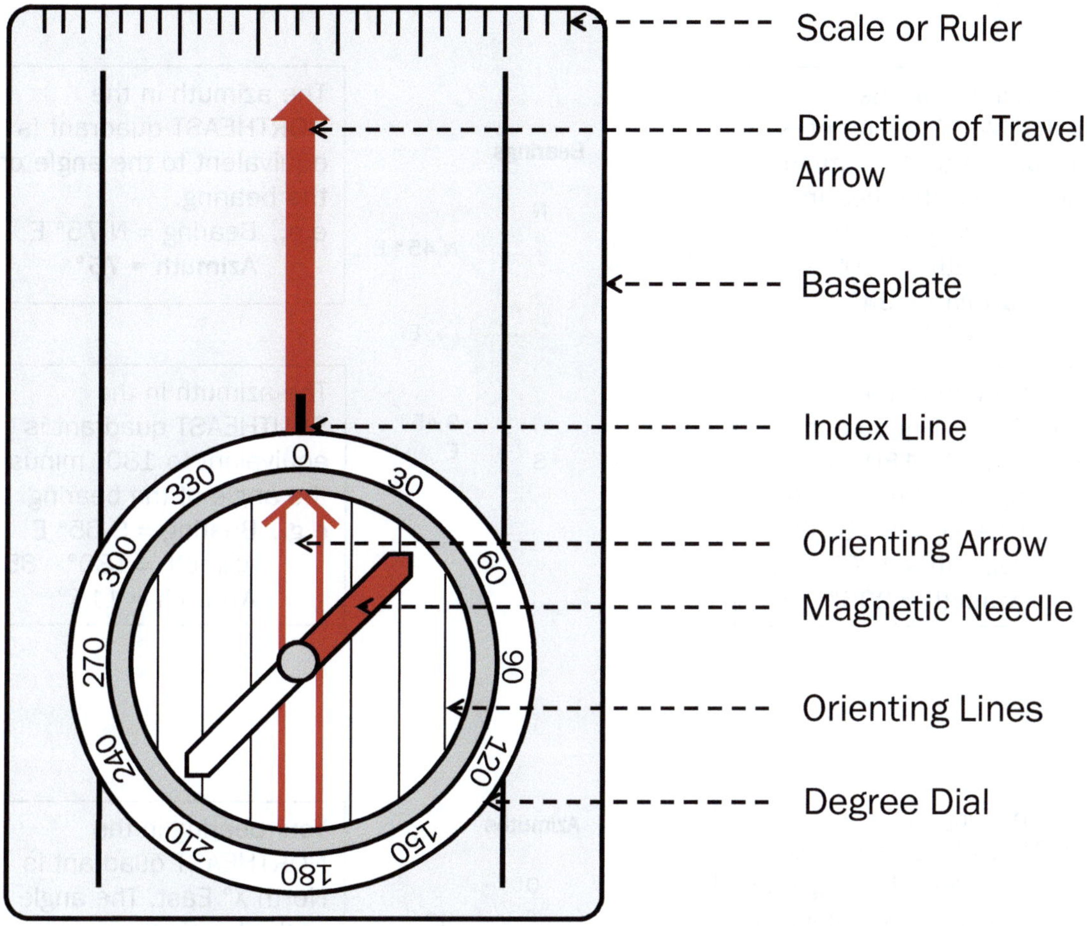

Bearings and Azimuths

A **bearing** in land navigation refers to the angle that describes the position of something in relation to a north–south line (i.e., meridian). For instance, you might be headed on a bearing of North 45° East, which would mean that your destination is 45° east of true north.

An **azimuth** is an angular measurement on a spherical coordinate system. In other words, when using an azimuth to describe the position of something, the angle is between 0 and 360°.

Converting between bearings and azimuths takes practice. Study the rules for converting between bearings and azimuths.

Converting bearings to azimuths

The azimuth in the NORTHWEST quadrant is equivalent to 360° minus the angle of the bearing.
e.g., Bearing = N 15° W
Azimuth = 360° – 15°
Azimuth = 345°

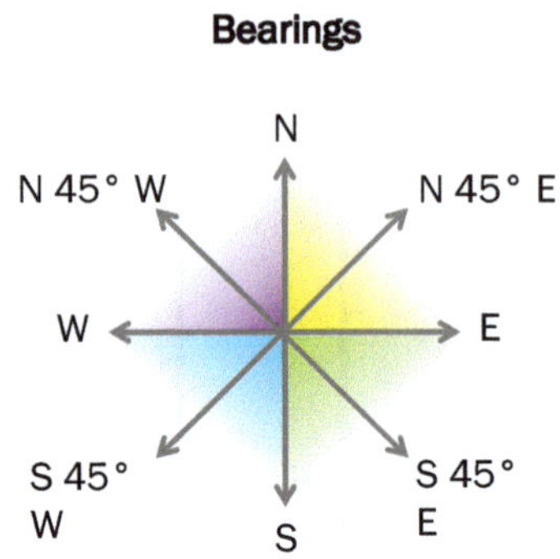

The azimuth in the NORTHEAST quadrant is equivalent to the angle of the bearing.
e.g., Bearing = N 75° E
Azimuth = 75°

The azimuth in the SOUTHWEST quadrant is equivalent to 180° plus the angle of the bearing.
e.g., Bearing = S 25° W
Azimuth = 180° + 25°
Azimuth = 205°

The azimuth in the SOUTHEAST quadrant is equivalent to 180° minus the angle of the bearing.
e.g., Bearing = S 65° E
Azimuth = 180° – 65°
Azimuth = 115°

Converting azimuths to bearings

Your bearing in the NORTHWEST quadrant is North X° West. The angle of the bearing is equivalent to 360° minus the azimuth.
e.g., Azimuth = 300°
Bearing = N 360° – 300° W
Bearing = N 60° W

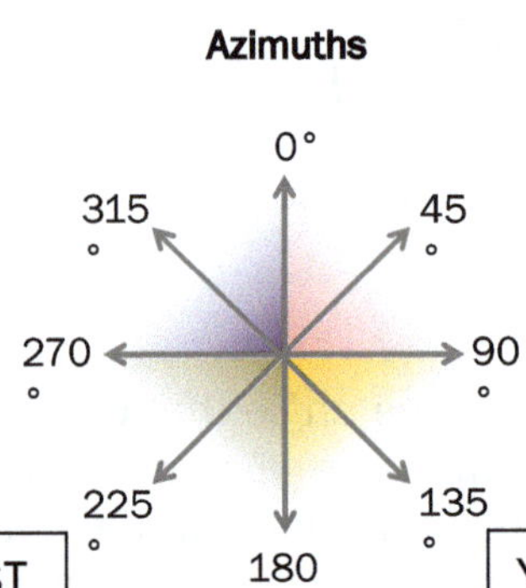

Your bearing in the NORTHEAST quadrant is North X° East. The angle of the bearing is equivalent to the azimuth.
e.g., Azimuth = 50°
Bearing = N 50° E

Your bearing in the SOUTHWEST quadrant is South X° West. The angle of the bearing is equivalent to the azimuth minus 180°.
e.g., Azimuth = 225°
Bearing = S 225°–180° W
Bearing = S 45° W

Your bearing in the SOUTHEAST quadrant is South X° East. The angle of the bearing is equivalent to 180° minus the azimuth.
e.g., Azimuth = 120°
Bearing = S 180° – 120° E
Bearing = S 60° E

Practice

1. What is the azimuthal equivalent to a bearing of S 34° E? ____________

► To solve this problem, it may be helpful to review the diagrams on the preceding page. The problem asks for an azimuth, so you know that the answer must be an angle between 0° and 360°. You also know that this direction is in the southeastern quadrant, which narrows the possible answers to an angle between 90° and 180°. Recall that an azimuth in the southeast quadrant is equal to 180° minus the angle of the bearing. So, the azimuthal equivalent of S 34° E is equal to 180° minus 34°. The answer is 146°.

2. What is the bearing equivalent to an azimuth of 321°? ____________

► Once again, review the previous diagrams to help you solve this problem. First determine the quadrant of the azimuth. An azimuth of 321° is in the northwest quadrant. So, the bearing equivalent will have the directions North and West. The next step is to determine the angle of the bearing. Bearings always have an angle from a North–South baseline. So the angle is equivalent to 360° minus the azimuth. The answer is 39°.

Compass Use: Determining Direction

Successfully using a compass requires skill and determination. You can use a compass in two ways. First, to determine a direction from a map or other source, set your compass to the desired azimuth in the field and navigate to your destination. Or, determine an azimuth of an object in the field from your current position. Of course, when navigating using maps and a compass together, you must consider **declination**, which is the difference between magnetic north and true north. You will consider declination during navigation in the next chapter.

To set your compass to a given azimuth, you simply turn the degree dial such that your given azimuth is indicated at the index line. Then, turn yourself and your compass so that the magnetic needle lines up with the orienting arrow. Whenever navigating, it is important to remember to *always keep the direction of travel arrow pointed toward your destination.*

► For example, pretend that you needed to navigate in a direction of 210°.

1. Turn the dial such that 210° is at the index line.
2. Turn yourself and the compass so that the orienting arrow lines up with the magnetic needle.
3. Proceed in the direction of travel until you reach your destination.

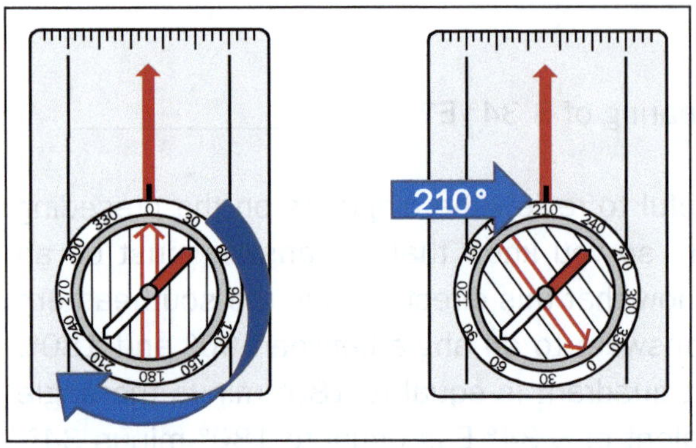

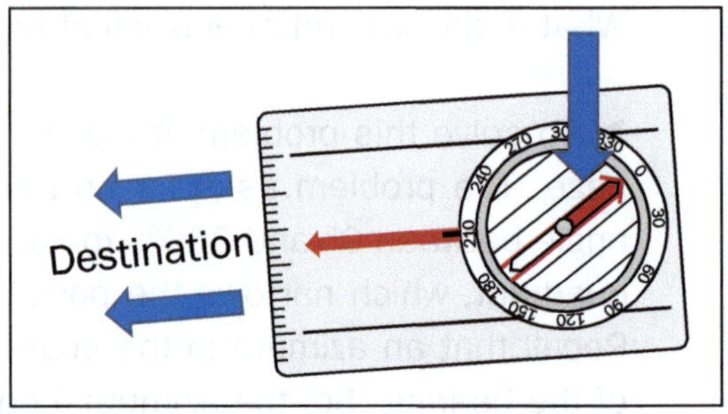

To determine an azimuth from your current position to an object in the field, point the direction of travel arrow toward the object. Next, turn the dial of the compass so that the orienting arrow lines up with the magnetic needle. Finally, read the azimuth at the index line.

► For example, pretend that you want to determine the azimuth from your current position to the tree ahead.

1. Point the direction of travel arrow toward the object.
2. Turn the dial so that the orienting arrow lines up with the magnetic needle.
3. Read the azimuth at the index line.

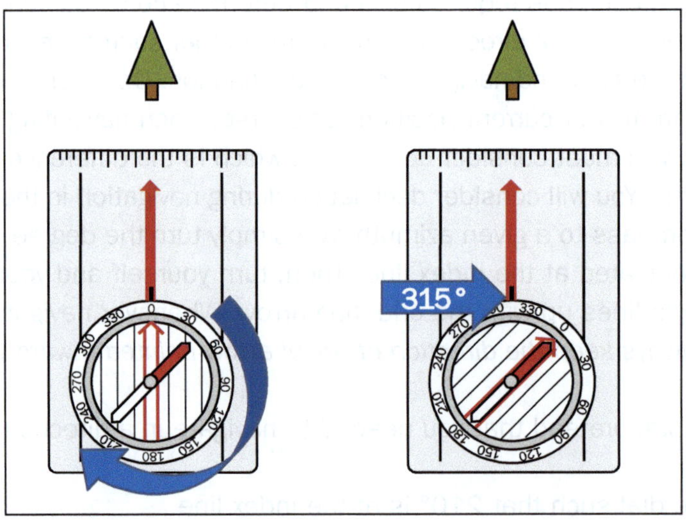

Pacing and Distances

Pacing is the method of measuring distance by using one's stride. One pace is the distance spanned with two steps (e.g., each time one's right foot hits the ground).

Source: http://pixabay.com/en/footprints-feet-walking-toes-human-149398/.

To determine your pace in units of distance per pace, it is useful to walk some known distances several times and *divide the known distance by the number of paces it took you to walk it*. It may also be helpful to determine how many paces it takes you to walk a specific distance, such as 100 ft. If your pace is slightly different for each distance, take the average pace. It might also be helpful to determine your pace on different terrains (uphill, over rocks, through mud, etc.).

Practice

1. If it took you 17 paces to walk 85 feet, how many feet would 1 pace be?

► The problem is asking you to determine the distance of 1 pace in units of feet. To solve this problem, simply divide the number of paces by the distance.

$$\frac{85\ feet}{17\ paces} = 5\ feet/pace$$

2. If your pace was 1.6 m, how many paces would it take you to travel 0.5 kilometers?

► First, find the number of meters in 0.5 km. There are 1,000 meters per kilometer. So 0.5 km equals 500 m. Next, divide 500 m by the distance of one pace:

$$\frac{500\ \cancel{m}}{1} * \frac{1\ pace}{1.6\ \cancel{m}} = \sim 312\ paces$$

SUMMARY

Navigation and orienteering are skills that involve determining position and direction using maps and compasses. There are several different models of compasses, but all of them have similar parts that are critical in determining the direction of travel. Directions may be measured using bearings or azimuths. Bearings refer to angles in relation to a north–south line, whereas azimuths use angular measurements between 0–360°. Successful orienteers understand pacing, which is a method of measuring distance by using one's stride. Understanding direction, distance, position, and the relationship between map and earth features will help you to navigate successfully.

TEST YOUR SKILLS

1. Determine your pace for 50 m (164 ft) and 100 ft (30.5 m). Compute the distance of your pace in units of m/pace and ft/pace.

2. Practice determining an azimuth from your position to an object in the distance. Pace the distance to the object and check your estimated distance with a tape measure.

3. Practice navigating to different flags or objects from a starting location, given specific azimuths and distances.

4. Convert the following bearings to azimuths:

    a. N 40° W ______________

    b. S 60° E ______________

    c. N 75° E ______________

    d. S 12°W ______________

5. Convert the following azimuths to bearings:

    a. 14° ______________

    b. 102° ______________

    c. 216° ______________

    d. 304° ______________

6. Solve the following:

    a. Joey's pace is 5 feet. How many paces would it take for him to walk 225 feet?

    b. Paul's pace is 1.5 m. How many paces would it take for him to walk 300 m?

    c. Susan's pace is 4.5 feet. She walked 97 paces. How far did she travel?

    d. Abby's pace is 1.2 m. She walked 102 paces. How far did she travel?

REFERENCE

Jacobson, C. 1999. *Basic Essentials: Map and Compass*, 2nd Edition. Guilford, CT: The Globe Pequot Press.

4

Navigation and Triangulation

Navigation is a skill of determining position and direction. There are several ways to determine your position and direction. For instance, a Global Positioning System (GPS) receiver can indicate your position and direction of travel; you can also use a map to determine your position. A compass or other navigation device can determine direction; shadows point west in the morning and east in the afternoon; star constellations circle the North Star.

Suppose, however, that you did not have a GPS. You only have a map and a compass, but are not sure where you are. You can determine your position by observing the features in the landscape that you can see and identifying them on a map. This technique is referred to as triangulation. **Triangulation** is the process of determining a location by measuring angles to that location from the location of known points. When using a map and compass together for triangulation or navigation, you must account for declination.

Declination is the difference between magnetic north and true north. A compass does not point to true north. It points to magnetic north, which is currently about 1,200 km (745 mi) southeast of the North Pole. Magnetic north changes over time due to continuous changes in the earth's magnetic field. The magnetic field is created, in part, from molten iron and nickel flowing in the earth's core. Declination is zero only along the **agonic line**. Areas west of the agonic line have an east declination. That is, if you are standing west of the agonic line, your compass will point too far east, and therefore must be corrected. Areas east of the agonic line have a west declination. The difference between magnetic north and true north presents a challenge when using a compass in conjunction with a map because maps indicate the direction of true north, whereas a compass points to magnetic north.

The purpose of this section is to help you refine your skills using a compass and reading maps, use a compass and maps together, account for declination, and triangulate your position on a map. Additionally, you will gain practice converting between several different units of scale that you likely will encounter during your career in natural resources.

LEARNING OBJECTIVES

1. Successfully navigate using a compass and map
2. Explain declination and adjust for it during navigation
3. Determine land distance using a map scale

4. Identify an unknown location using triangulation
5. Explain the relationship between maps and the earth

Determining Direction and Distance from a Map

Many natural resources–related jobs require field skills that include navigation. Therefore, it is imperative to understand how to find your way using maps and a compass. Several skills are needed for successful navigation. First, you need to know how a compass works. Next, you need to understand how to read map features. Third, you need to understand map scale to convert between map distance and earth distance. Finally, you need to know your declination. In this section, you will learn to determine direction from a map using your compass and a straight edge.

Suppose you are viewing a map and would like to travel from point A to point B.

Steps to determine direction from a map

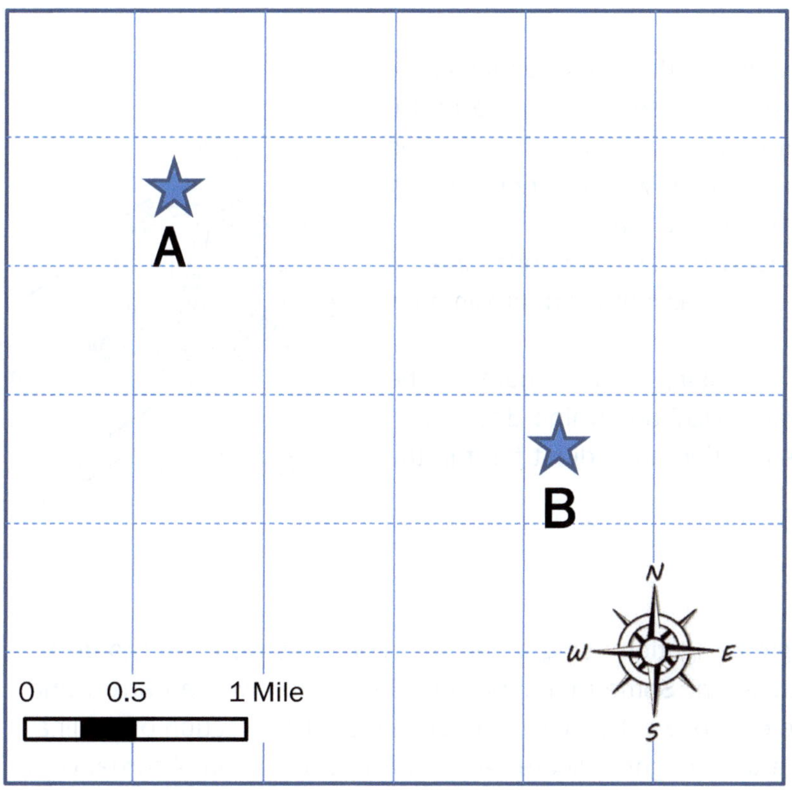

1. Using a straight edge, draw a line between A and B. Then, orient the compass so the direction of travel arrow is pointing from A toward B.

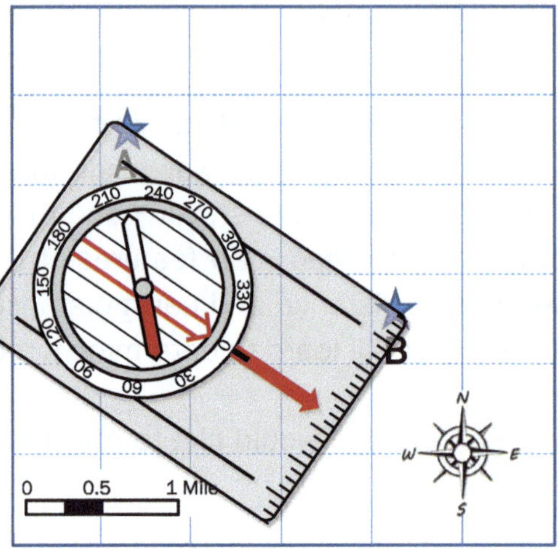

2. Turn the dial so that the orienting arrow on the dial is lined up with grid north. In other words, make sure the orienting arrow is parallel with the north–south meridians on the map.
3. Read your azimuth at the index line. In this case, the azimuth is approximately 120°.
4. If you are using this azimuth in the field, you MUST determine declination for this location and adjust for it in the field.

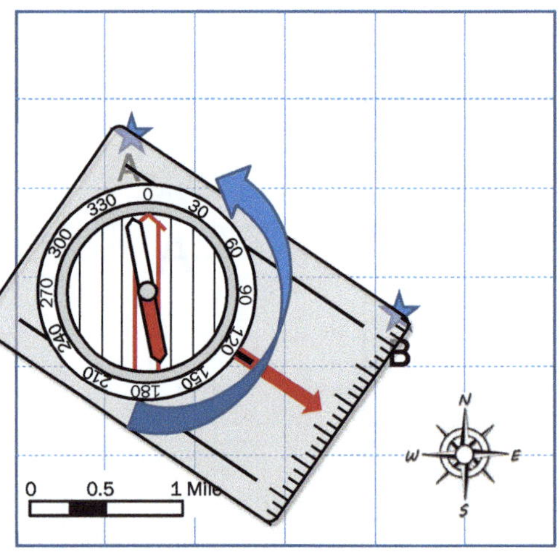

Declination

Before you begin navigation using a map and compass, you need to determine your declination. Some maps indicate declination with an arrow and an angle to the east or west of north, depending on the direction of declination. The image to the right indicates a 10° EAST declination. Alternatively, you can determine declination using an **isogonic chart** (below) or compute your declination for a specific location through online magnetic field calculators such as the one through the National Oceanic and Atmospheric Administration: http://www.ngdc.noaa.gov/geomag-web/#declination.

Using the following isogonic chart, first locate the agonic line. Recall that declination is zero along the agonic line. Next determine the approximate declination at your current location.

► ____________ degrees EAST or WEST (circle one)

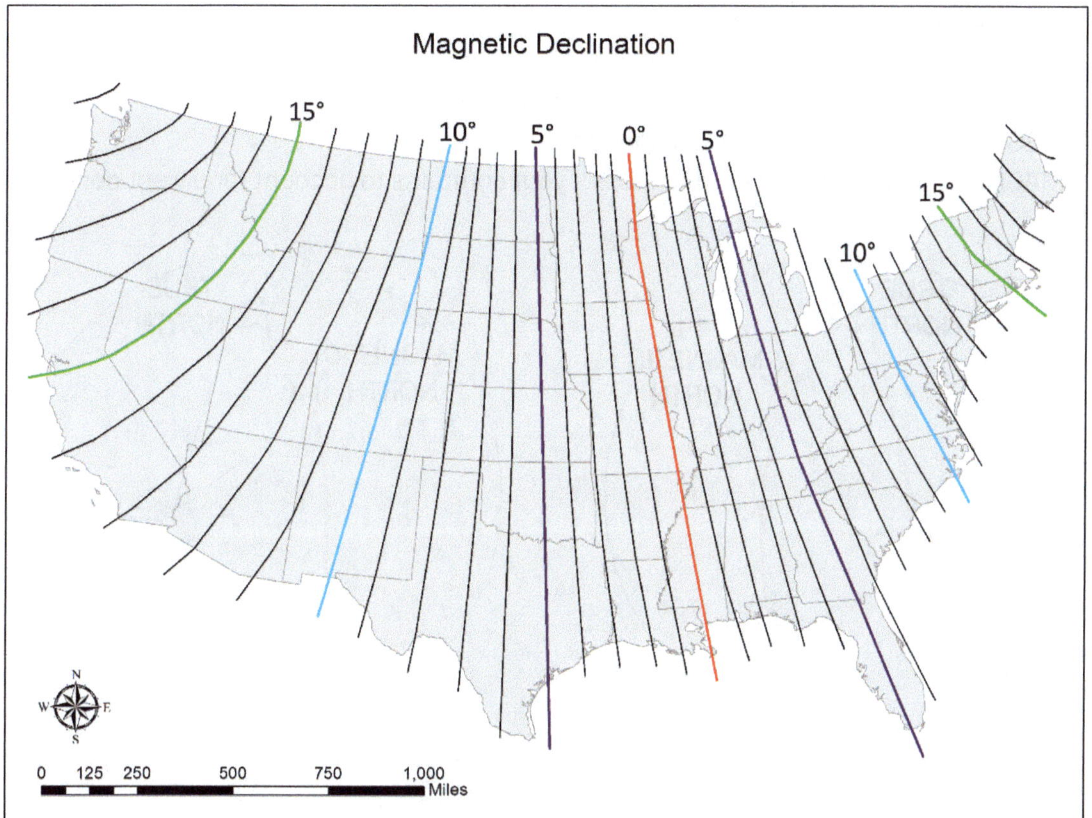

"United States Magnetic Declination 2005," http://www.ngdc.noaa.gov/geomag-web/, U.S. Department of Commerce.

Areas west of the agonic line have an east declination. The compass needle points too far east because it is attracted to magnetic north, which is south of true north. Thus, when setting your compass azimuth to a reading you took off a map, you need to SUBTRACT the angle of declination from the true azimuth determined from the map. When marking an azimuth on a map from a reading you took off your compass in the field, you do the opposite. A convenient way to remember how to adjust for declination is the following equation:

TRUE – EAST = MAGNETIC

where,

TRUE = the true azimuth determined from a map
EAST = the angle of an east declination
MAGNETIC = the azimuth at which to set your compass to account for an east declination

Likewise, in areas east of the agonic line, a compass needle points too far west. Thus, when setting your compass to an azimuth determined from a map, you need to ADD declination. When marking an azimuth on a map from a reading you took off your compass in the field, you do the opposite. A convenient way to remember how to adjust for declination is the following equation:

TRUE + WEST = MAGNETIC

where,

TRUE = the true azimuth determined from a map
WEST = the angle of a west declination
MAGNETIC = the azimuth at which to set your compass to account for a west declination

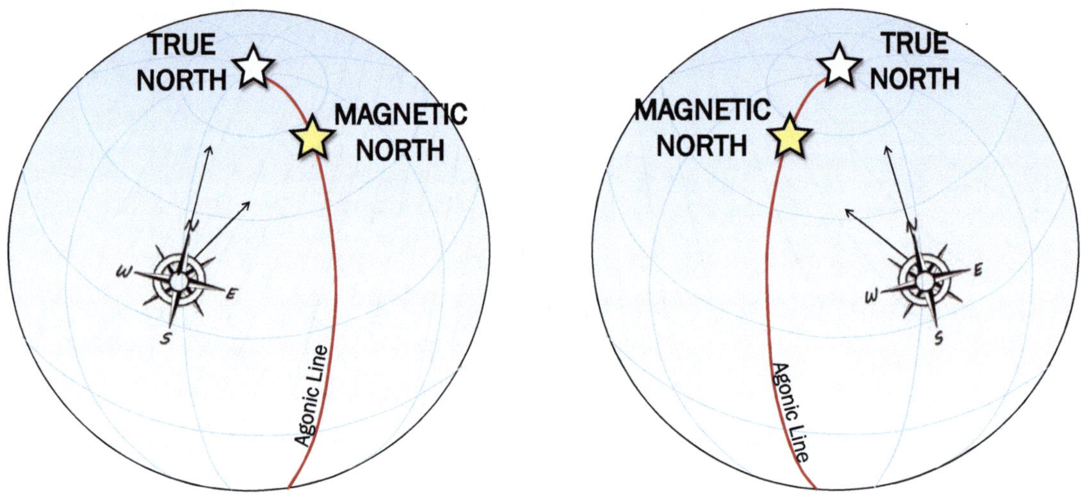

So, suppose you were in a geographic area with a declination of 5° WEST. If a map indicated that you needed to travel on an azimuth of 87° to reach your destination, on what azimuth would you need to set your compass?

► ____________° TRUE + WEST = MAGNETIC; solve for MAGNETIC

In the same geographic area, if you were in the field and determined that a unique tree was 214° from your current position, what azimuth would you use to mark its direction on a map?

► ____________° TRUE + WEST = MAGNETIC; solve for TRUE

Suppose you were hiking in Idaho and knew you had a declination of 15° EAST. If you used a map to determine that your next camping destination was on an azimuth of 104° from your current position, on what azimuth would you need to set your compass to reach your destination?

► ____________° TRUE – EAST = MAGNETIC; solve for MAGNETIC

In the same geographic area, if you were in the field and determined that an amazing rock formation was 315° from your current position, what azimuth would you use to mark its direction on a map?

► ____________° TRUE – EAST = MAGNETIC; solve for TRUE

Practice

► Suppose the geographic area represented in the map below is in Allendale, Michigan. The declination in Allendale is 5° West. At what azimuth would you set your compass if you wanted to travel:

1. From A to B? ____________________
2. From A to C? ____________________
3. From A to D? ____________________
4. From C to D? ____________________

► Using the map scale, determine the distance between:

1. A and B ____________________
2. A and C ____________________
3. B and D ____________________
4. C and D ____________________

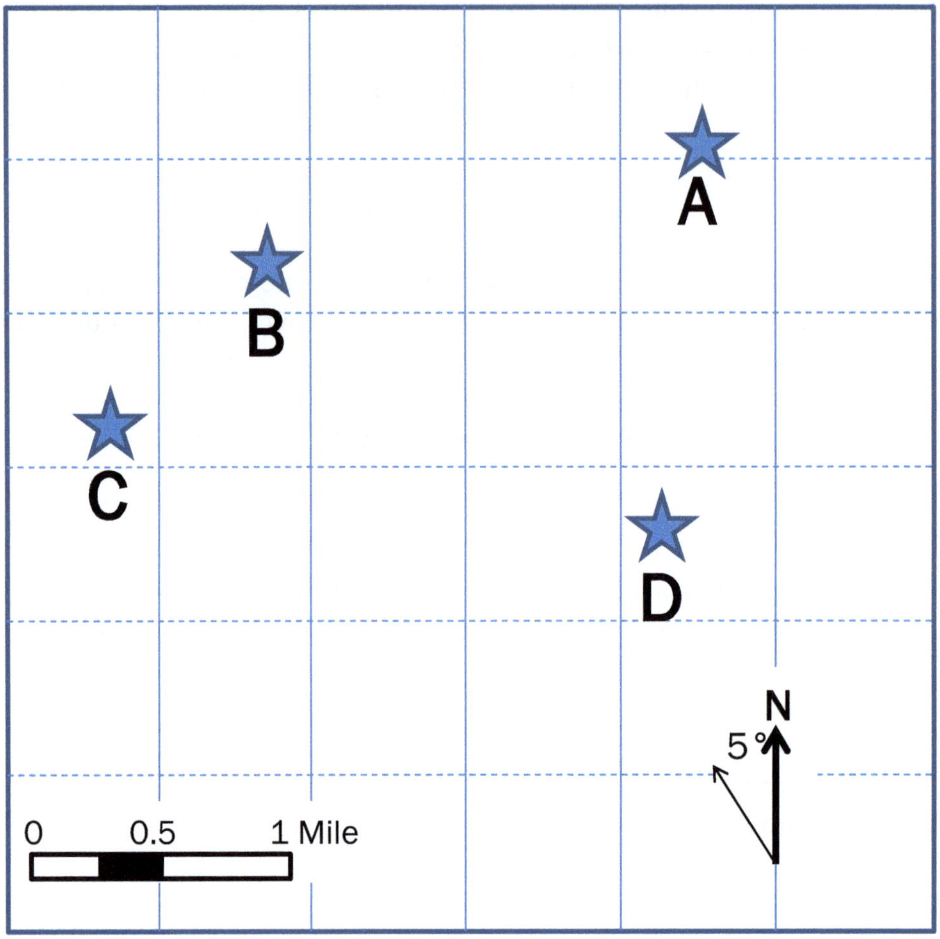

Triangulation

Triangulation is a great skill to learn. In the field, you may use triangulation to accurately mark the position of an object on a map. For instance, if you are tracking an animal, you may use triangulation to locate its position. Perhaps you are bass fishing in a very productive spot and decide you would like to mark its location on a map so you can return to it at a later time. **Triangulation** is the process of identifying the location of a point by determining the angle between it and the location of at least 3 points with known position. For example, pretend that you want to determine your exact position on the college campus in the following map. Using a compass in the field, you determined that the azimuth from you to the center of the Life Science building is 25°, the azimuth to the northwest corner of the Law building is 95°, and the azimuth to the northeast corner of the Forestry building is 128°. You know from the information on the map that in this geographic area, declination is 6° East. Where are you?

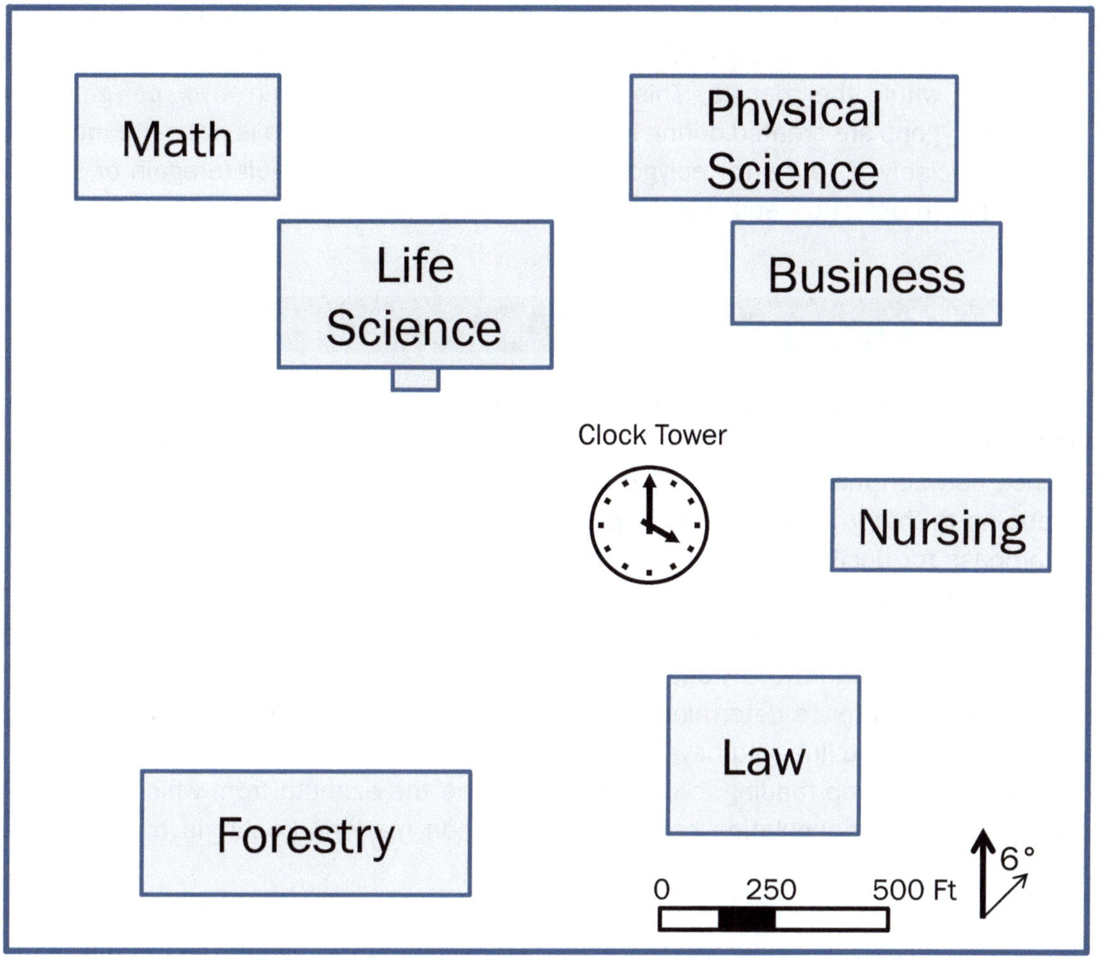

Steps to practice triangulation

1. In the field, position yourself in a location where you can see three objects or features that are also evident from your map.
2. WITHOUT MOVING from your position, take an azimuth reading from your position to the first object.
3. Correct for declination. Using the map, draw a line along the azimuth extending from the object. You are standing somewhere along the line.
4. WITHOUT MOVING from your position, take an azimuth reading from your position to the second object.
5. Repeat step 3. You are likely standing where the two lines intersect, but a third line would confirm your position.
6. WITHOUT MOVING from your position, take an azimuth reading from your position to the third object.

7. Repeat step 3. You are standing where all three lines intersect. If you have created a little triangle with the intersections of all three lines, you are standing within the triangle. This triangle is referred to as an **error polygon**. Error polygons are created during triangulation when each angle is not quite measured precisely. If your error polygon is large, you should triangulate again or select a fourth object to reduce the margin of error in your positioning.

SUMMARY

Navigation using a map and compass is an important skill in many field-related jobs. When using a compass and map to navigate, it is important to consider declination, which is the difference between magnetic north and true north. Because a compass needle points to magnetic north, but grid north on a map approximates true north, knowing how to adjust your compass for declination will help you to navigate correctly. Isogonic charts or online declination calculators will help you to compute declination for specific geographic areas. On the agonic line, declination is 0° because magnetic and true north align. If you are east of the agonic line, you have a west declination and must add the angle of declination to a map reading in order to determine the azimuth from which to set your compass. If you are west of the agonic line, you have an east declination and need to subtract the angle of declination from a map reading in order to determine the azimuth from which to set your compass. Skills in triangulation come in handy when marking positions of features on maps.

TEST YOUR SKILLS

1. Have someone hide a treasure and mark its location on a map. Using the map and the features on it, triangulate your exact position. Then, use the map to acquire an azimuth (don't forget to adjust for declination). Use the map scale to estimate the distance from your current location to the treasure. Go find it!
2. Please answer the following:

   a. What is the azimuthal equivalent to a bearing of N 80° E? ______________

   b. What is the azimuthal equivalent to a bearing of S 60° W? ______________

   c. What is the bearing equivalent to an azimuth of 200°? ______________

   d. What is the bearing equivalent to an azimuth of 320°? ______________

e. If you walked from your truck into the woods to your study area on an azimuth of 290°, on what azimuth would you set your compass when you want to return to your truck?

f. Using the map to the right, what is the azimuth from point A to B?

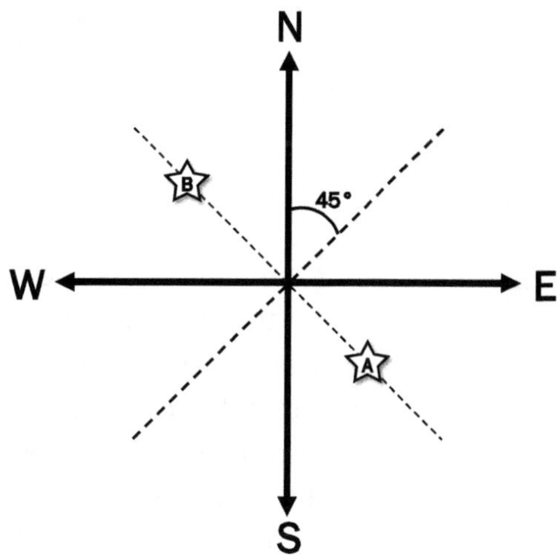

g. Using the same map, what is the azimuth from B to A?

h. Using the same map, if you had a declination of 10° W, at what azimuth would you set your compass in the field to travel from A to B?

i. Using the same map, if your declination was 4° E, at what azimuth would you set your compass in the field to travel from A to B?

REFERENCES

Campbell, J. 2001. *Map Use & Analysis*, 4th Edition. New York, NY: McGraw-Hill Higher Education.

Jacobson, C. 1999. *Basic Essentials; Map and Compass*, 2nd Edition. Guilford, CT: The Globe Pequot Press.

Wikipedia: Triangulation http://en.wikipedia.org/wiki/Triangulation

5

Introduction to Topographic Map Reading

Topography is the description of the shape and elevation of features on the surface of the earth. **Topographic maps** are used to depict elevation and shape of land features. They are different from **planimetric maps**, which depict only the horizontal position of features.

In previous chapters, you learned how to use a compass between maps and the field, estimate distance by pacing, account for declination, and successfully navigate to a specific destination. In this chapter, you will continue refining your navigation and map-reading skills and you will begin to understand how topographic features are represented on maps.

LEARNING OBJECTIVES

1. Interpret land features on a topographic map
2. Determine land distance using a map scale and topographic information
3. Triangulate your position using topographic features
4. Calculate slope

Understanding Topographic Features on Maps

On topographic maps, **contour lines** connect points of equal elevation. Thus, lines spaced closely together indicate a steeper elevation than lines spaced further apart. You gain or lose elevation as you travel from one contour line to another. If you are traveling along a contour line, you will not be changing elevation. The **contour interval** is the vertical distance between contour lines. For instance, if one contour indicated 100 feet above sea level and the next contour indicated 150 feet above sea level, the contour interval is 50 feet.

Recognizing specific patterns in contour lines will help you visualize landscapes. Below are common features to help you begin to recognize them on topographic maps. Indicate each topographic feature on the map.

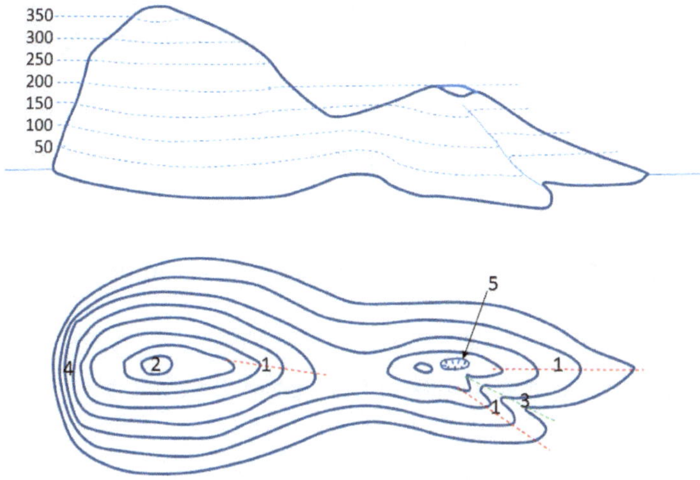

1. Ridgeline or Spur
2. Peak or summit
3. Valley, gulley, or draw
4. Cliff
5. Depression

A **ridgeline** or **spur** occurs when land descends gradually from a higher elevation to a lower elevation. On topographic maps, ridgelines or spurs are indicated by contour lines that are drawn into a U-shape, and the bottom of the U points to a lower elevation. A **peak** or **summit** is the highest point of elevation. On a contour map, a peak is found where there are no other contour lines to indicate a higher elevation immediately adjacent to it. **Valleys, gullies,** or **draws** are formed between ridges on either side. These areas can be formed from erosion or washout. The contour lines that represent these features generally form a U- or V-shape, where the bottom of the U or V points to a higher elevation. A cliff occurs where there is a steep change in elevation within a relatively short distance. Adjacent contour lines are drawn close to each other to indicate a steep elevation change. A **topographic depression** is the bottom of a bowl, or an area surrounded by higher elevation in all directions. On a topographic map, depressions are indicated with hatched contour lines so that they are distinguished from peaks.

PRACTICE

► Use the map of Lost Lake to determine the following:

1. Is the river flowing into or out of Lost Lake? How can you tell?
2. What might challenge you on a canoe trip from point A to Lost Lake?
3. You are standing at point C. What is your elevation?
4. What is your elevation at point D?

5. What is the true azimuth from point B to point C (measured off the map)? If you were in the field, at what azimuth would your compass need to be set to account for declination?

6. You are standing at point C. Describe the topography ahead if you oriented yourself at 297° in the field.

7. What is your azimuth from point C to point D? If you walked this azimuth, would you gain or lose elevation?

8. Describe the terrain you would cover if you walked from point C to point E.

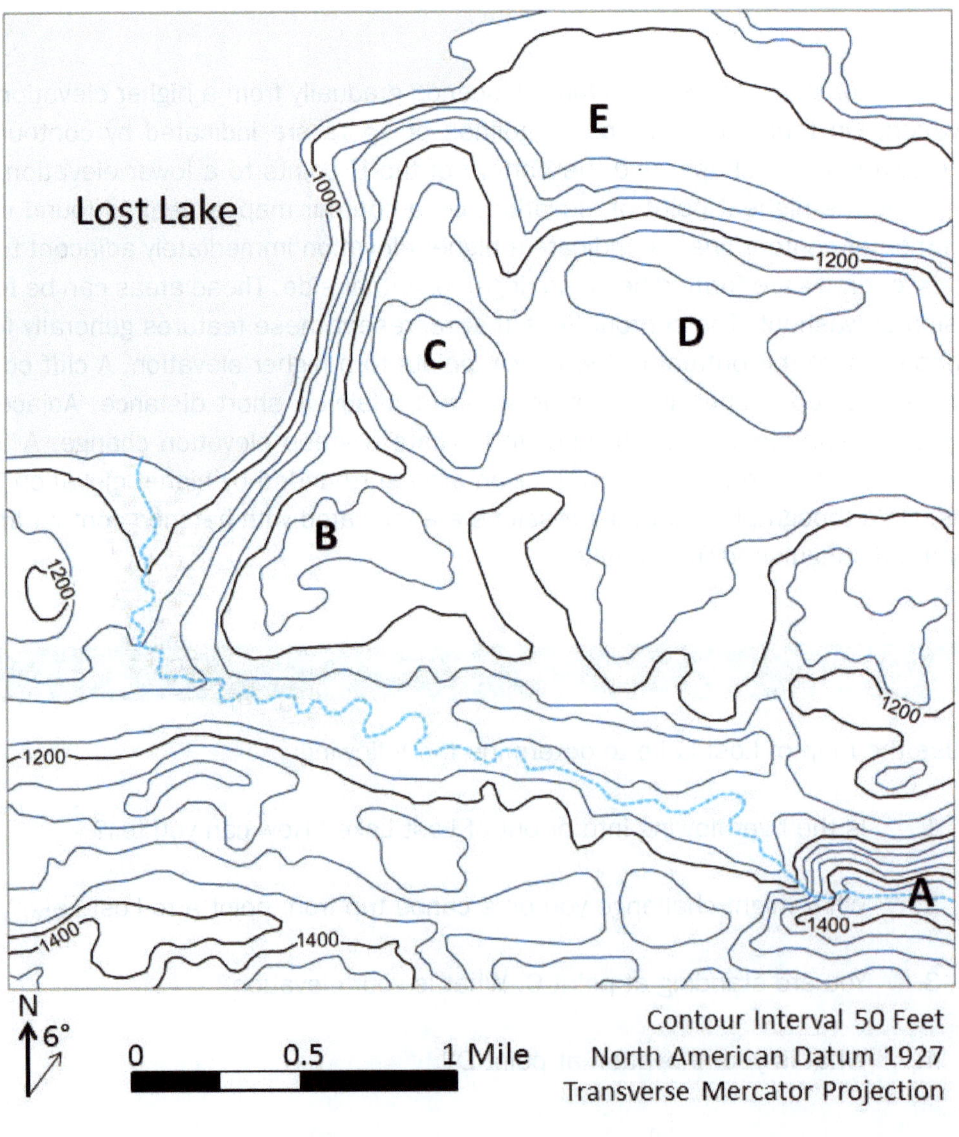

Distance Calculations with Topography

You have used scale on maps to determine distance on the ground, but your previous calculations were on relatively flat surfaces. If you are in a hilly or mountainous terrain, distance calculations are a bit more involved, because you need to consider slope.

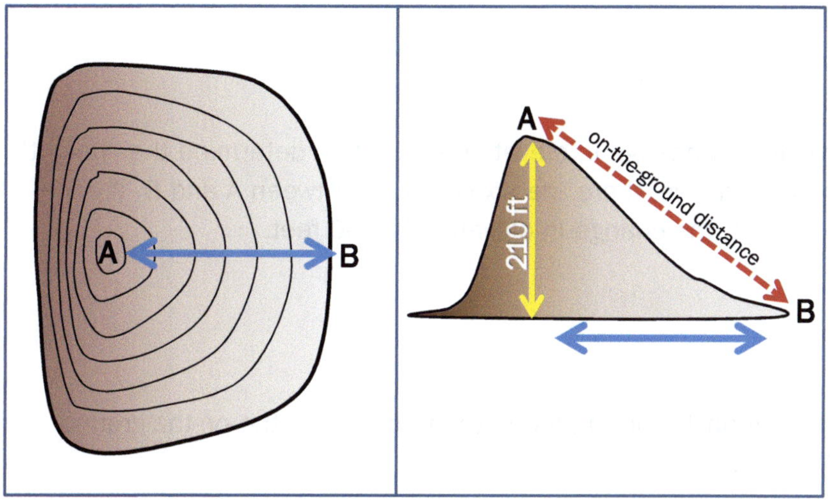

► In the map below, suppose the measured map distance between A and B is 1 inch. Also, suppose each contour represents 30 feet in elevation. What is the on-the-ground distance between A and B?

Step 1. Represent a cross-section of the land intersecting points A and B as a right triangle. Measure the map distance between A and B. This distance is the "base" of the triangle.

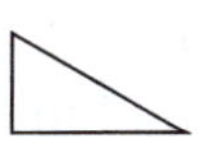

1 inch

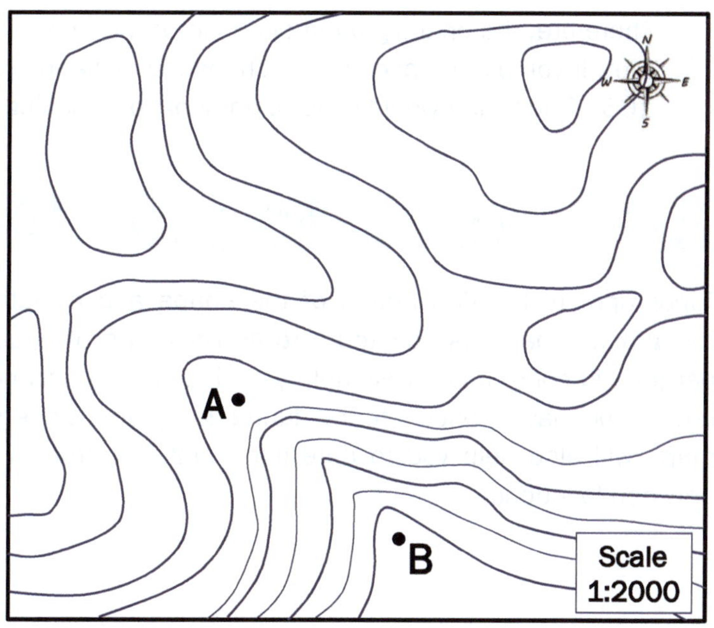

Step 2. Using the scale, convert the map distance to earth distance in a relevant unit (e.g., feet).

Thus, if 1 inch on the map is equivalent to 2,000 inches on the ground, then 2,000 inches = 166.67 ft.

So, the base of the triangle is 166.67 feet on the ground.

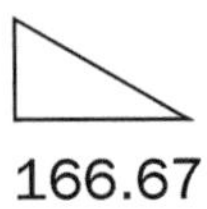

Step 3. Determine what your change in elevation is. That is, determine the "height" of the triangle. In the map above, there are six contours between A and B. If each contour represents 30 feet, the change in elevation is 180 feet.

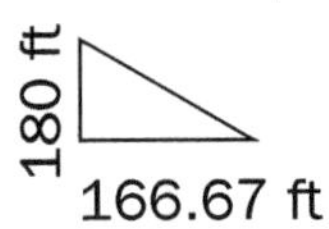

Step 4. Use the Pythagorean Theorem (geometry) to calculate the on-the-ground distance between A and B.

$a^2 + b^2 = c^2$

$180^2 + 166.67^2 = c^2$

$c = 245.3$ ft

Therefore, the on-the-ground distance between A and B is approximately 245 feet. If you did not account for change in elevation, you would have only traveled 166.67 feet and would have been short of reaching your destination.

SUMMARY

Topography is the description of the shape and elevation of the earth's surface. Many topographic maps use contours to depict areas of equal elevation. Recognizing specific patterns in contour lines will help you visualize landscapes and understand what types of terrain you may be encountering during navigation. Understanding how to read topographic maps will also help you to determine distance from a map if elevation changes occur between locations.

TEST YOUR SKILLS

Please use the topographic map below to solve the following questions:

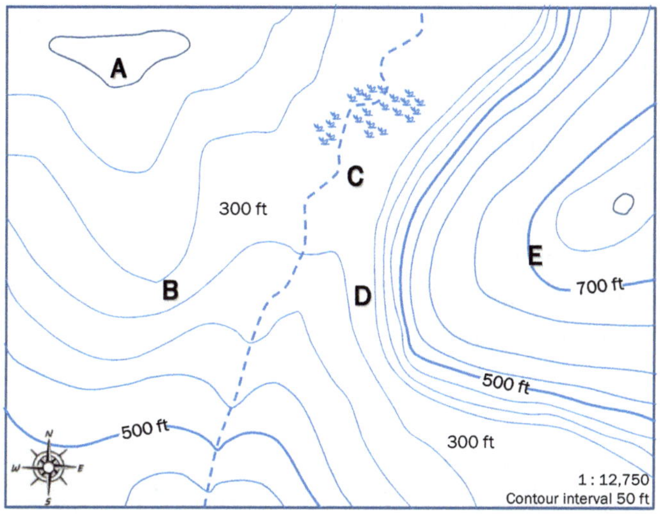

1. What is the elevation at point A?
2. What is the elevation difference between points A and B?
3. Describe the topography if you walked from A to B.
4. If the distance between C and D measures 0.81 inches on the map, what is the earth distance between them in units of feet and miles?
5. If the distance between D and E measures 1.125 inches on the map, what is the on-the-ground distance between them in units of feet and miles?

REFERENCE

Jacobson, C. 1999. *Basic Essentials: Map and Compass*, 2nd Edition. Guilford, CT: The Globe Pequot Press.

6

Boundary Surveys and Plats

Surveying is the technique and science of accurately determining the position of points and the distances and angles between them. You may have seen surveyors on the side of a road under construction or perhaps you have had your property boundaries surveyed. Surveys may be conducted for property boundaries or for topography. **Boundary surveys** are conducted by Professional Land Surveyors to establish the corners of a parcel of land and identify its location using a **legal description**. Land parcels are recorded in **plat maps**: maps showing the ownership divisions of a piece of land. In this chapter, you will learn about the general process of surveying, but more importantly, you will learn how to read a plat map, determine legal descriptions of land parcels, and calculate measurements.

LEARNING OBJECTIVES

1. Define *boundary surveying*
2. Define *plat map*
3. Determine legal descriptions from plat maps
4. Determine area and distances from plat maps

Boundary Surveying

Professional licensed surveyors measure distances, angles, and elevation, and record precise locations of property corners. Surveying tools may include rods, levels, transits or **theodolites** for measuring angles, global positioning systems, or **total stations** (i.e., devices containing an electronic theodolite and electromagnetic distance-measuring instrument). Essentially, surveyors identify the position of a feature on the parcel, such as a corner, and mark it with a monument. A **monument** is a physical object secured in the ground to mark survey points and lines. Surveyors continue to identify and mark locations of features along the property boundary and carefully record the information. Surveyors produce detailed records of the boundary survey, create plat maps, and register their records and maps with the county in an office related to property titles, deeds, or surveys.

Surveyors using a total station (right) and prism (left) to determine distances and angles between boundary corners.

Survey monuments

Public Land Survey System

The **Public Land Survey System** (PLSS) is a rectangular survey method that describes the location of land parcels in the United States (except for Maine, New Hampshire, Vermont, Massachusetts, Rhode Island, Connecticut, New York, New Jersey, Maryland, Virginia, North Carolina, South Carolina, Georgia, Tennessee, Kentucky, Texas, and parts of Ohio). In a rectangular survey method, land is divided into townships of essentially equal size. **Townships** are land parcels that are 36 square miles. They are subdivided into **sections** of 1 square mile, and may be further divided into fractions of sections. The states listed above use a different survey method (the European method, using metes and bounds), whereby each land parcel is individually described and bounded using landmarks and compass bearings. For instance, a boundary may have been described using metes and bounds as 35° east of north from the large oak stump, thence following the bank of Bubbly Stream. The metes and bounds method may result in uncertainty about exact boundaries as landmarks changed. The PLSS is a more straightforward method of describing land parcels.

The PLSS is essentially a coordinate system. Locations of land parcels may be described by directions in reference to north–south **Principal Meridians** or east–west **Base Lines**. Have you ever driven down a road called Meridian Road, or one called Baseline Road? The names of these roads are significant within the PLSS. Townships are numbered sequentially and labeled with a north–south or east–west direction in relation to the origin of the intersection between a Principal Meridian and Base Line. Numbers north or south of a Base Line are called **tiers**, and numbers east or west of a meridian are called **ranges**. For example, Township 2 North, Range 1 East refers to the position of a township that is two tiers NORTH of the Base Line and the first range EAST of the Principal Meridian.

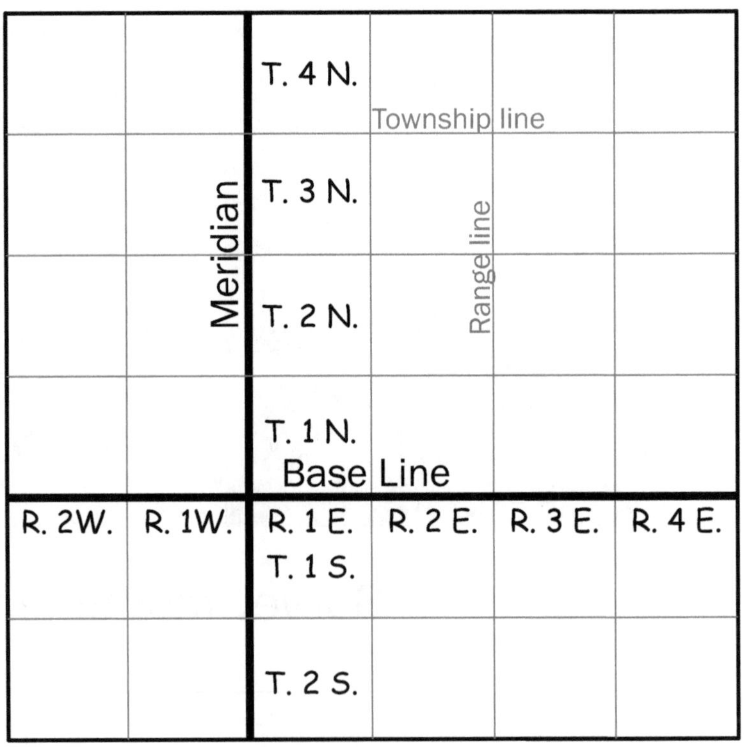

Townships are divided into thirty-six sections that are one square mile each. Sections are numbered sequentially starting from the northeastern corner of the township.

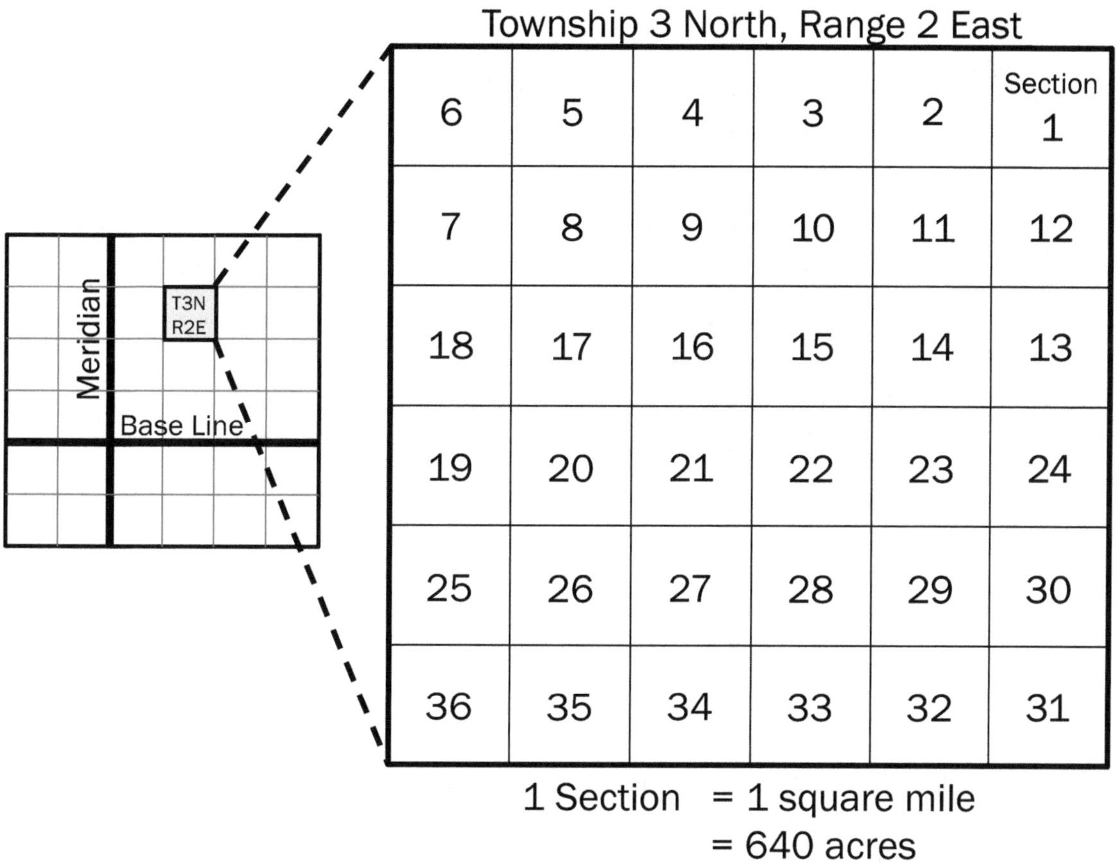

Legal Description and Area

From the rectangular surveys under the PLSS, land area and location descriptions are clearly understood. Plat maps illustrate the divisions of land parcels based on township, ranges, and sections. Legal descriptions refer to the written location of a property. A written legal description of a land parcel begins with the smallest fraction of land within a section and ends with the township and range. For instance, the legal description for the 320-acre parcel in the diagram below is the S ½, Sec 23, T 3N, R 2E. In words, the parcel is the southern half of Section 23, in Township 3 North, Range 2 East. The 10-acre parcel of land identified by the star in the diagram below is NE ¼, NW ¼, NW ¼, Sec 23, T 3N, R 2E. The parcel is the northeast quarter of the northwest quarter of the northwest quarter of section 23, in Township 3 North, Range 1 East.

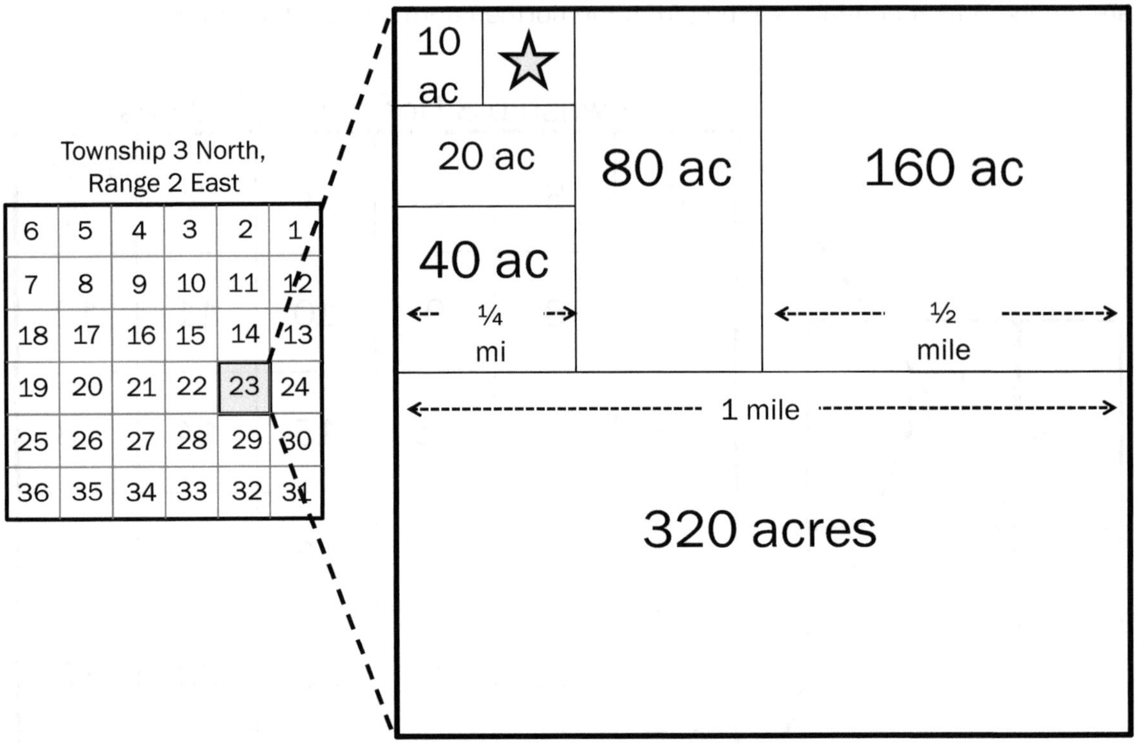

Area and length measurements are easily determined as fractions of sections. For example, you know that one section is 1 square mile, or 640 acres. To determine area, simply divide by the number of divisions there are in the section. In other words, if the section is divided by 2, each land parcel is 320 acres. If the section is divided by 8, each land parcel is 80 acres. Length is computed in a similar fashion. Each side of a section is 1 mile. To determine length and area of irregularly shaped land parcels, principles of geometry or map scaling tools may be used.

Divisions from the PLSS rectangular surveys are evident when viewing the landscape with aerial imagery. In many areas, roads run in a north–south or east–west direction along section boundaries. In the infrared image below, land parcels are depicted clearly through patterns of roads and land use. In natural resources management fields, it is helpful to be able to estimate area and length of land parcels by observing sections and fractions of sections created by patterns of land use within the landscape.

Infrared aerial image of landscape.
Sections and sub-divisions are evident.

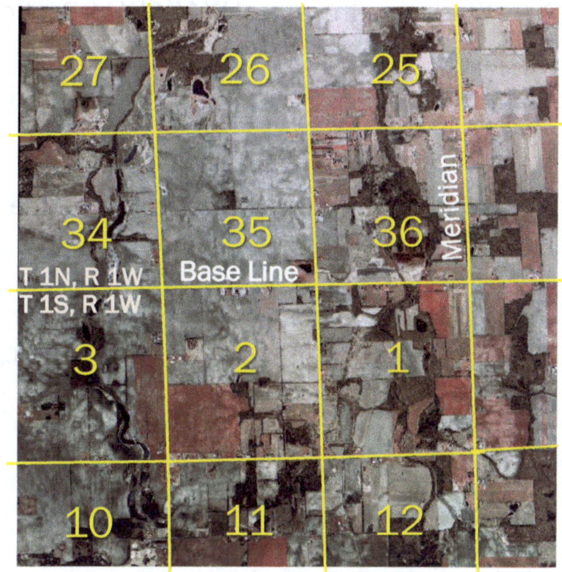

Source: U.S. Department of Agriculture.

SUMMARY

Surveying is a science and technique that accurately determines the position of points and the distances and angles between them. Boundary surveying is used to determine legal descriptions of land parcels. The Public Land Survey System (PLSS) is a rectangular survey system that is widely used in the United States. This system replaced a survey method called metes and bounds whereby land parcels were individually described and bounded using landmarks and compass bearings. The PLSS uses rectangular divisions referred to as townships, ranges, and sections. A township is 6 square miles and runs north or south of a base line. Ranges run east and west of a principal meridian. Sections are 1-square-mile divisions within townships. The PLSS is used in the majority of the United States to identify the locations and boundaries of land parcels and record them using legal descriptions.

TEST YOUR SKILLS

Using the Sheridan Township plat map below, please answer the following questions:

1. Who owns the parcel S ½ Section 15 T 15N R7W? ______________________________

2. Who owns the parcel SE ¼ NE ¼ Section 32 T15N R7W? ___________________________

3. Who owns the parcel W ½ SW ¼ Section 20 T15N R7W? ____________________

4. What is the legal description of the parcel owned by F & E Lyons in section 21?
____

5. What is the legal description of the parcel owned by F. Cressey in section 9?
____

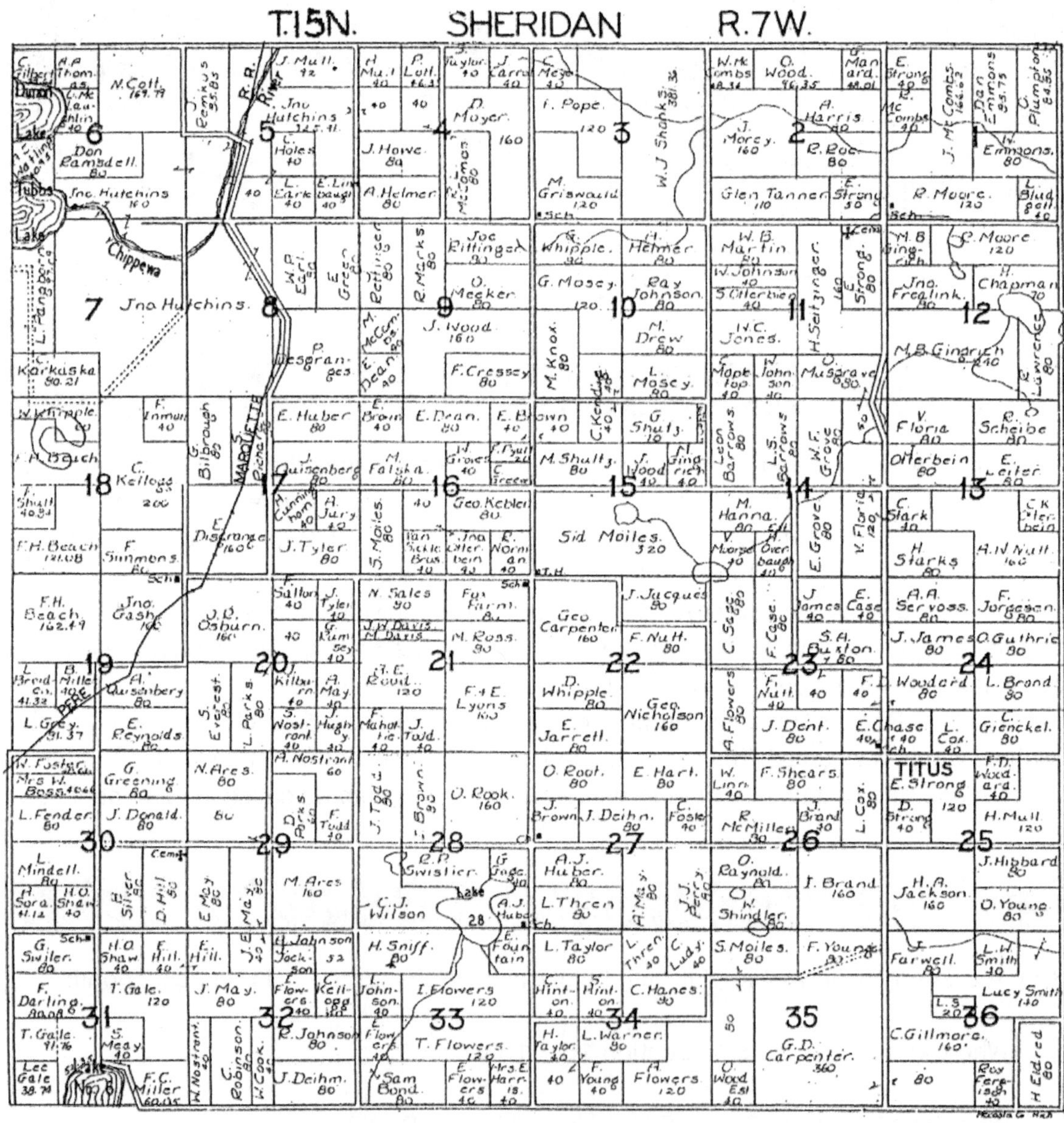

Source: http://www.migenweb.org/mecosta/1918platmaps/1918She.htm.

6. What is the legal description of the parcel owned by O. Raynold in section 26?
____

7. In section 10, how many acres does M. Drew own? ____________________

8. In section 11, how many acres does W. Johnson own?
____

9. In section 29, what is the E–W distance across M. Ares' parcel?
____

10. In section 9, what is the perimeter length of J. Wood's parcel?
____

REFERENCES

Avery, T. E. & Burkhart, H. E. (2002). *Forest Measurements*, 5th Edition. New York, NY: McGraw-Hill.

Gay, P. (2014). *Practical Boundary Surveying: Legal and Technical Principles*. New York, NY: Springer.

Neunzert, G. M. (2010). *Subdividing the Land: Metes and Bounds and Rectangular Survey Systems*. Boca Raton, FL: CRC Press.

USGS. "The National Map." http://nationalmap.gov/small_scale/a_plss.html

7

Topographic Surveying

A **topographic survey** contains information depicting the three-dimensional position of land features and the distance and angles between them. Natural resources–related professionals use topographic surveys to create maps depicting elevation and contours. Often, topographic surveys are used to analyze landscape features, measure stream channel characteristics, create contour maps, identify water flow potential, evaluate erosion risk, etc. Sometimes topographic surveys are compared for multiple years to assess the change in land over time. The process of collecting elevation data along line transects or within a target area is called **differential leveling**. This chapter will familiarize you with survey equipment, the process of surveying, data organization, and graphical depiction of an elevation profile.

LEARNING OBJECTIVES

1. Define *leveling*
2. Explain the process of differential leveling
3. Use topographic data to create a contour map
4. Graph a profile of the elevation change along a transect

Leveling

Leveling is essentially the process of measuring differences in elevation. **Indirect leveling** involves calculation of elevation from slopes and horizontal distances. Indirect leveling may be done from topographic maps. **Direct leveling** is the process of measuring differences in elevation directly. Two methods of direct leveling are profile leveling, where measurements are taken along a line transect to create an elevation profile, and contour surveying, where elevations are found at points within a grid over a target area.

Survey Equipment

Several instruments are important when conducting a topographic survey. We'll discuss a few of them here. A **survey rod** or **leveling rod** is a graduated rod used to measure the vertical distance between a point on the ground and the line of sight of a surveyor's level. Rods are 7–9 feet in length and can be extended to approximately twice their length. The face of the rod is marked in feet and subdivided into hundredths of a foot. A **laser level** is a device containing a laser beam projector with a rotating head containing a mirror for sweeping the laser beam about a vertical axis. The level is supported on a tripod. As the laser level spins around, the laser hits a **target**, which is affixed to a survey rod and signals when the laser is detected. Elevation measurements can be determined by taking a reading on the rod when the target signals. When surveying a transit, a **compass** is handy for keeping direction and position accurate. Finally, a tape measure is important for measuring distance along a transect.

Survey Process

The first step in conducting a topographic survey is to identify the boundaries of your target area. Once the area is identified, you can determine the location of transects or, alternatively, identify a sampling intensity. Where will you record elevation? How frequently?

Second, you must establish a reference point or a benchmark. The **benchmark** is a position with a known elevation (e.g., distance above sea level) or one to which the elevation of all measurements will be referenced (e.g., 100 feet). A benchmark should be a secure feature that will not likely move if subsequent surveys are to be taken. During the process of differential leveling, it is not imperative that the actual elevation of the benchmark be known. The difference in elevation between points is more important than the actual elevation.

Third, set up a laser level or other leveling device. The level will be affixed to the tripod and set within view of the benchmark *and* within view of the target area. The position will typically be at a high elevation within the target area. An automatic level is self-leveling, so once the tripod is set, the laser will

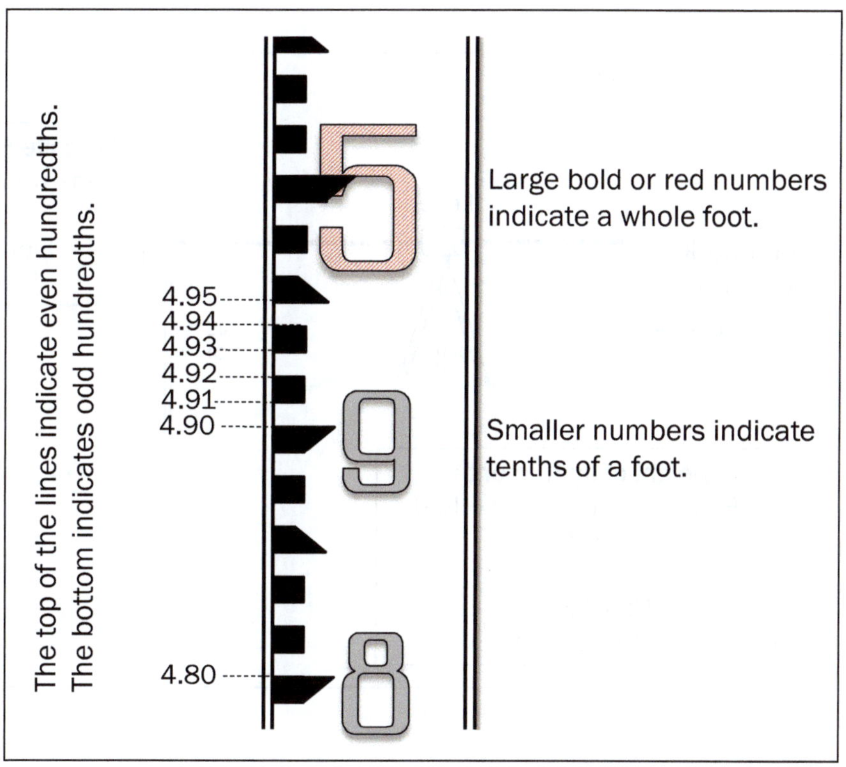

automatically adjust itself to be perpendicular to a vertical plane. The level and tripod collectively are referred to as the base station.

Fourth, if you are using a laser level, you must secure the target on the leveling rod. Then, the rod must be positioned at the benchmark. A measurement at the benchmark or at a position of known elevation is referred to as a **backsite**. One person must hold the leveling rod completely vertical and raise the rod until the target sounds a constant tone. A second person must read the number at the index line on the target. If you are not using a laser level, you should look through your level to read the rod. This first reading plus the benchmark elevation will give you the height of the instrument (i.e., base station). The height of the instrument will be used to calculate the elevation of all other sample points.

Fifth, once you have the height of the base station, you are ready to take more elevations along a transect to complete your sample. These subsequent elevation measurements are referred to as **foresite** elevations. The foresite elevation can be determined only after a difference is calculated between the backsite and the foresite. Use your compass to ensure you are traveling along the proper bearing. Use the tape measure to measure distances between sample points. Record elevations at predetermined intervals or when there are significant changes in elevation. Repeat elevation readings at all sample locations within the target area. A turning point may be necessary if you need to move the level to a different location. If this is the case, then a point with known elevation (i.e., a former foresite with known elevation) can be used as a benchmark.

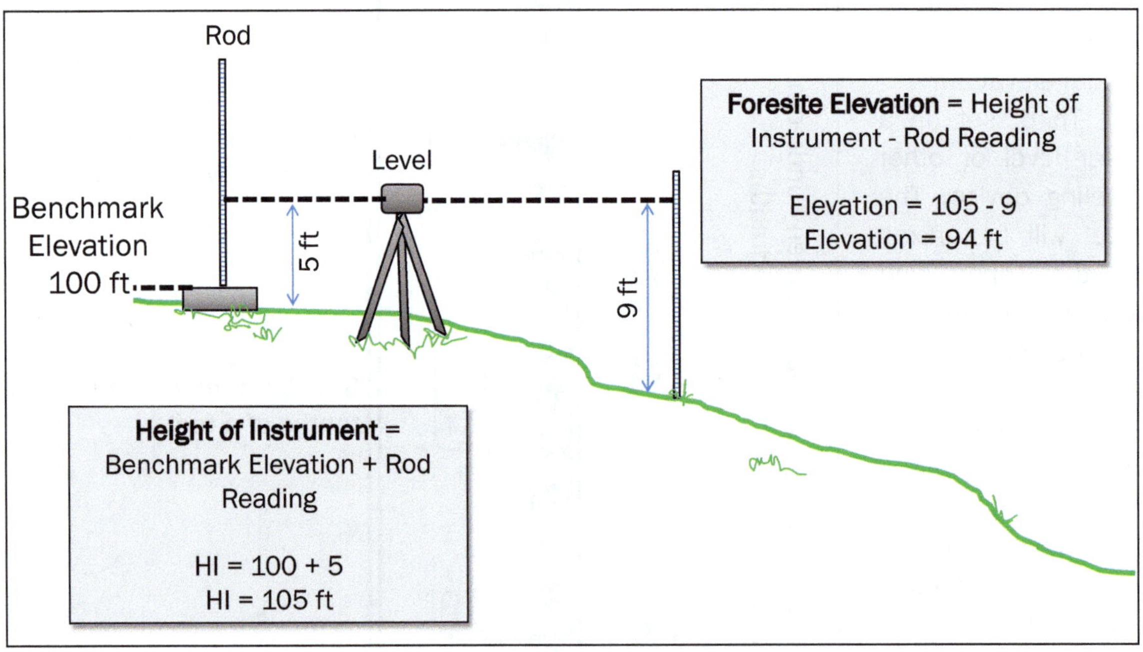

Below is an example datasheet that may be used for topographic surveys.

Transect: A

Date: 18-Sep-2015

Team: Ali, Heather, Paul, and Todd

Base station height: 4.72

Benchmark elevation: 100 ft

Height of instrument: 100 + 4.72 = 104.72

Distance from base station	Rod reading	Elevation	Notes
0	7.96	96.76	grassy area
10	8.01	96.71	sidewalk
20	8.05	96.67	grassy area
22	9.16	95.56	abrupt downward slope
23	9.3	95.42	land levels off
30	9.32	95.4	grassy area

Elevation = instrument height – rod reading

Using Topographic Data

Topographic survey data can be used to create topographic profiles or contour maps. A **topographic profile** is a diagram of the cross-section of the survey area. For instance, each survey point is plotted on a graph where the x-axis indicates distance along a transect, and the y-axis indicates elevation. You can calculate percent slope from profiles by dividing the change in vertical distance by the horizontal distance. For example, in the diagram below, the percent slope is

$$\frac{change\ in\ vertical\ distance}{change\ in\ horizontal\ distance} * 100$$

or, (40 m ÷ 100 m) * 100 = 40%. Thus, the slope of the land is relatively steep.

To compute slope in degrees, simple trigonometry may be used:

$$\tan(\theta) = \frac{change\ in\ vertical\ distance}{change\ in\ horizontal\ distance}$$

or,

$$\theta = tan^{-1}\left(\frac{change\ in\ vertical\ distance}{change\ in\ horizontal\ distance}\right)$$

In essence, the slope in degrees is 21.8°.

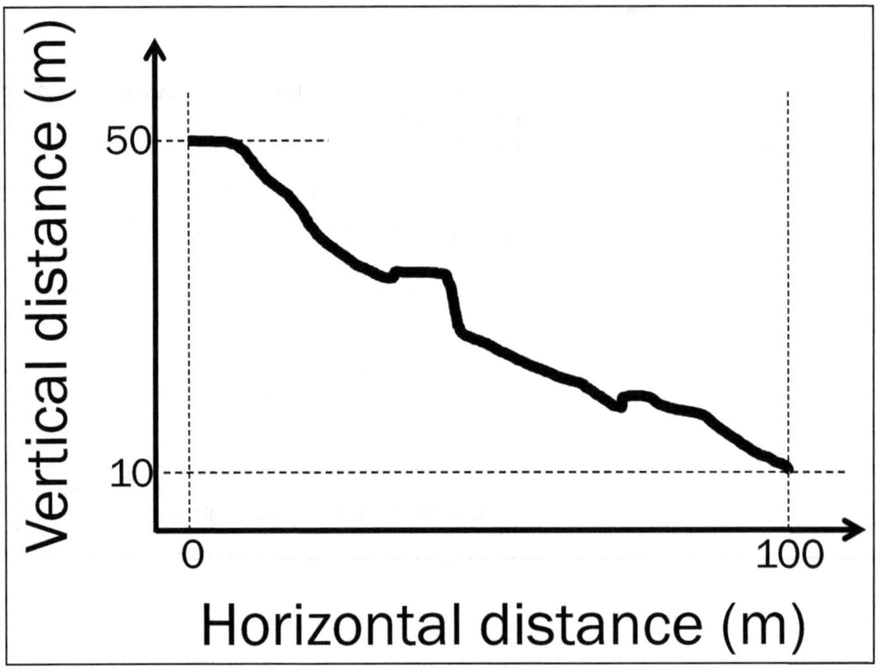

The horizontal distance is not the same as a surface or on-the-ground distance. A horizontal distance is perpendicular to the vertical distance (or elevation change). Surveyors can measure horizontal distances using chains or measuring devices that are kept horizontal relative to the base station. If the slope is relatively steep, elevation should be measured at smaller horizontal distances. Horizontal distance may be measured with a steel or fiberglass tape to prevent sagging of the tape measure and, therefore, inaccurate distance measurements. Electronic equipment such as total stations or Electronic Distance Measuring (EDM) devices may provide more sophisticated methods using lasers or other electromagnetic waves to measure horizontal distances precisely. Alternatively, if distance were paced or otherwise measured on the ground, horizontal distance may be computed using the Pythagorean Theorem:

$$a^2 + b^2 = c^2$$

or

$$b = \sqrt{c^2 - a^2}$$

Where the a is the height reading from the rod, b is the horizontal distance between rod positions, and c is the surface or on-the-ground distance between rod positions.

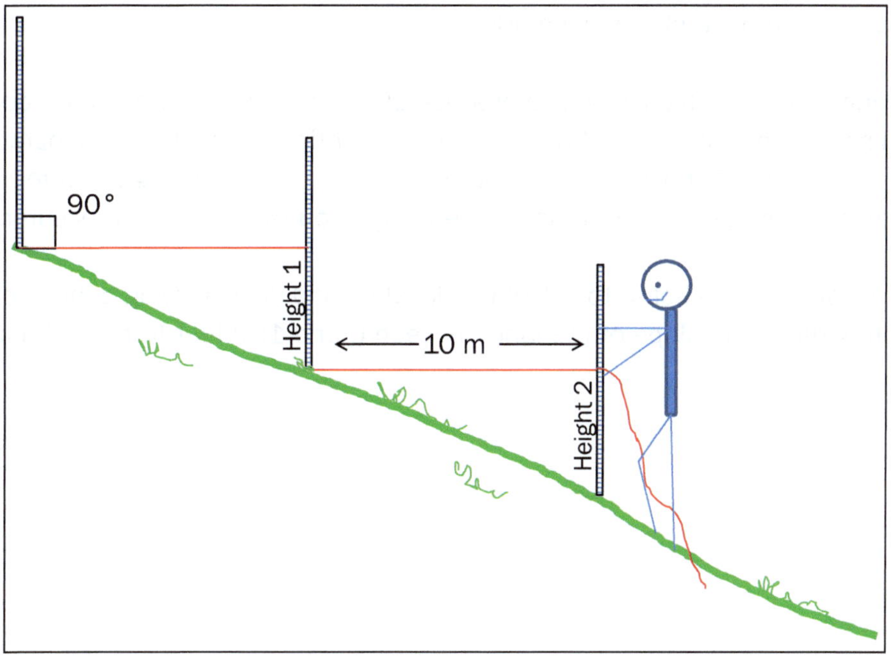

Topographic profiles may be used to study changes in topography over time. For example, soil conservationists may graph topographic profiles of land adjacent to streams each year to understand rates of erosion. Slope may be computed to determine how the gradient along the stream bank changes. **Contour maps** also indicate changes in elevation, but they represent an area versus a line transect. You will use example survey data in the next chapter to create contour maps.

SUMMARY

Topographic surveying is used to collect necessary information for creating topographic maps and analyzing the earth's terrain. Leveling is the process of measuring differences in elevation. Topographic surveyors establish a benchmark, or an elevation reference point, and then use survey rods, laser levels, and targets to measure differences in elevation along transects. The elevation measurements may be graphed to create topographic profiles, which can be used to calculate slope or study changes in topography over time.

TEST YOUR SKILLS

1. Explain the purpose of a benchmark.

2. Leveling is a cost-effective way to produce elevation data. How accurate do you think leveling is using various methods? Is accuracy critical in creating a topographic map? Why or why not? Can you think of instances where accuracy might matter more than in other circumstances? What measures might you take to ensure accuracy?

3. Use the survey data below to calculate elevation and draw a topographic profile on the graph. What is the slope of the land between 0 and 150 feet from the base station?

Transect: A

Date: 21-May

Team: Ali and Paul

Base station height: 5.12 ft   Benchmark elevation: 100 ft

Height of instrument: 105.12 ft

Distance from base station	Rod reading	Elevation	Notes
0	2.15		Grassy slope
10	5.49		
20	5.83		
21	5.10		Slight rocky outcrop
23	5.23		Same rocky outcrop
23	5.23		Same rocky outcrop
25	7.48		Edge of rocky outcrop
30	7.52		
40	8.03		
50	8.84		
53	8.21		Slight increase in elevation
55	8.94		
60	9.62		
70	9.87		
80	10.00		
90	10.43		
100	11.21		
110	11.78		
120	12.46		
130	12.87		
140	13.04		Bare soil from some apparent erosion
150	15.10		Edge of pond
190	15.10		Edge of pond
200	13.20		Bare soil from erosion
210	12.41		Grass and vegetation
220	11.98		
225	11.01		

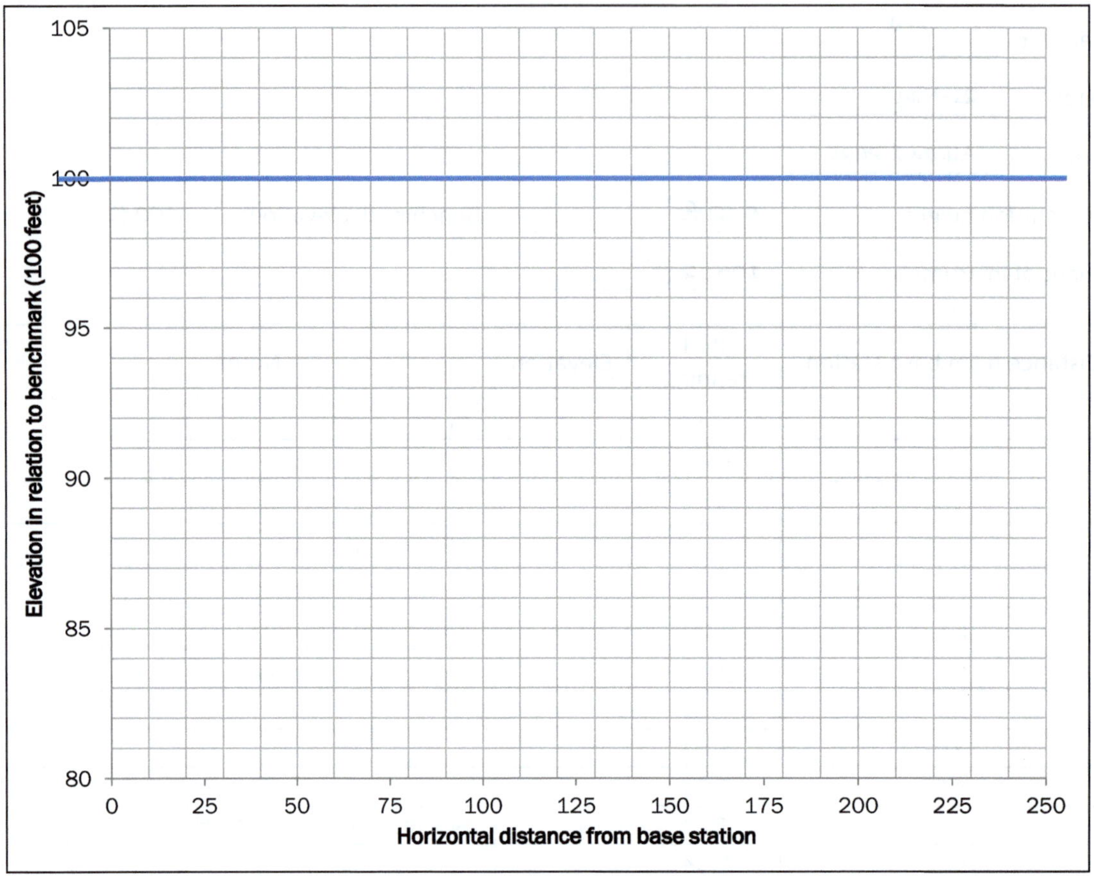

REFERENCES

Singh, R., Artman, D., Taylor, D. W., & Brinton, D. (2000). "Basic Surveying: Theory and Practice." Ninth Annual Seminar Presented by the Oregon Department of Transportation Geometronics Unit. Bend, OR. http://www.oregon.gov/odot/hwy/geometronics/docs/basicmanual2000_02.pdf

Wood, Ivan D. (1937). "EC758 Topographic Leveling and Preparation of Topographic Maps for Irrigation Purposes". Historical Materials from University of Nebraska-Lincoln Extension. Paper 2266. http://digitalcommons.unl.edu/extensionhist/2266

8

Understanding Topography and Watersheds, and Creating Surface Maps

In Chapter 7, you learned how a surveying technique called leveling is used to collect elevation data. In this chapter, you will learn how elevation data may be used to create surface maps depicting topography. A **surface map** is a two-dimensional representation of a three-dimensional surface. For instance, contour maps are surface maps using lines to depict the shape and elevation of land. Such maps may be used in natural resources professions to show landscape features, slope and aspect, or water flow.

Contour maps are useful in natural resources–related fields in several ways. You have used a contour map to determine distance down a slope. You have also used contour maps to begin understanding how the shape of contour lines depicts topographic features such as ridges, valleys, depressions, and summits. In this chapter, you will use contour maps to delineate watersheds. A **watershed** is the area of land where all the water under it or draining off it goes into the same place. In other words, a watershed defines the area of land where surface water would drain and converge to a single body of water at a lower elevation. Watersheds are important ecological management units, as they are a logical unit for studying water flow and hydrologic cycles. They can help resource managers understand overland flow, vegetation growth patterns, pollution, sedimentation, and nutrient flow.

LEARNING OBJECTIVES

1. Define *watershed*
2. Delineate watersheds
3. Understand characteristics of topographic maps and how to interpret them
4. Successfully create a contour topographic map
5. Successfully create a digital surface topographic map
6. Use the surface plot to visualize the topography and answer questions about the landscape
7. Understand resolution

Creation and Analysis of Contour Maps

Recall that topographic surveying (i.e., leveling) may be used to collect elevation data within a target area. Suppose in a specific target area, elevation measurements were taken at 10-foot intervals. A compilation of all survey data would produce a grid of elevations with a 10 x 10-foot resolution. **Resolution**, in this case, refers to the amount of detail in a grid. For instance, if we sampled every 5 feet, we would have produced a dataset with a finer resolution. If we used only the elevation measurements at 20-foot intervals, our sample resolution would be coarser. The amount of resolution in your dataset should depend on your sampling objectives. Sampling resolution within small areas is typically finer than in larger areas.

To create a contour map, simply connect points of equal elevation. Contour lines are smooth; they do not have sharp corners. Also, contour lines need to occur for every elevation indicated. In other words, if elevation drops sharply, contour lines should be close together. In the diagrams below, suppose that each number indicates a surveyed elevation. The resolution in the diagrams below is 10 x 10 feet. Thus, each square represents a 10-ft x 10-ft area.

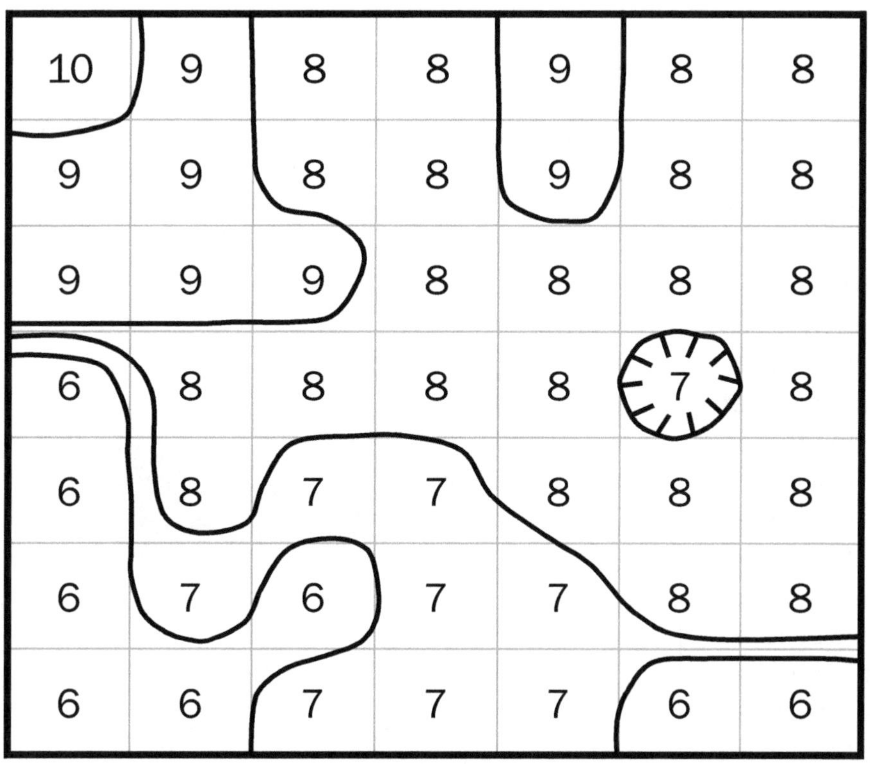

CORRECT Contour Map

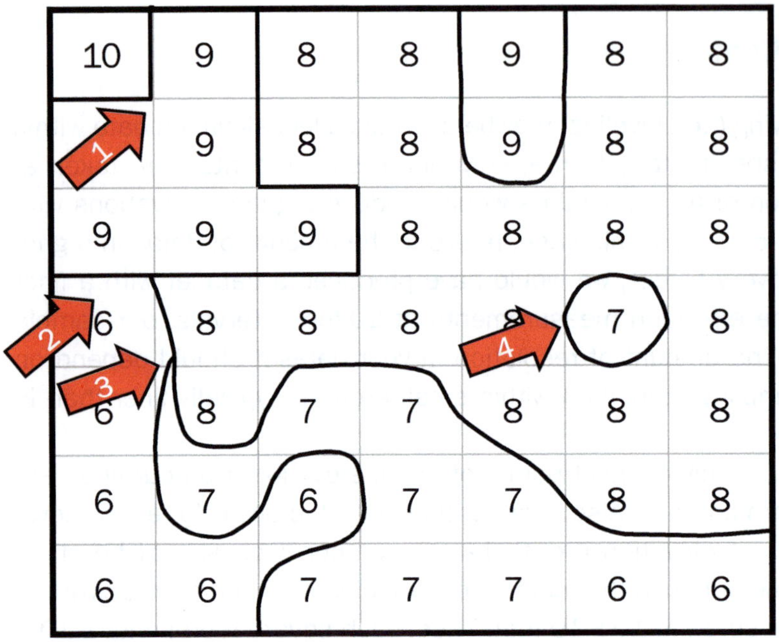

1. Contours are smooth. They do not make right angles.
2. A contour line must be represented for every elevation interval.
3. Contour lines do not intersect.
4. Depressions must be indicated as such.

INCORRECT Contour Map

Creating Digital Surface Maps Using Computer Software

Many computer software programs can be used to create digital surface maps using topographic data. Surface maps can even be created using common programs such as Excel. These types of surface plots can help you visualize the landscape in three dimensions, which often is easier to interpret and understand than trying to visualize a three-dimensional landscape from two-dimensional contours. The surface plot is an example of representing a landscape in 3D using plotted data from a topographic survey.

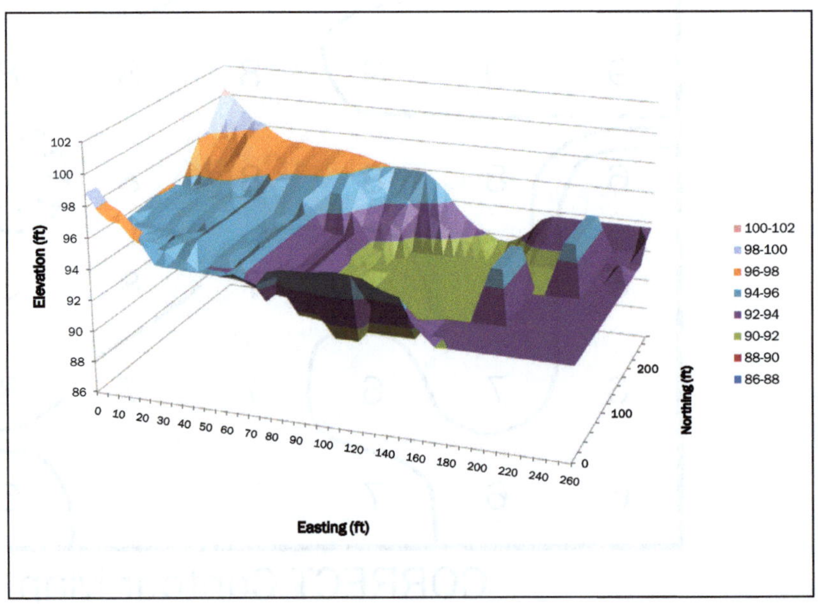

What Is a Watershed?

A watershed is an ecological management unit where all the water falling within it drains into a common location. Think of a basin. If water fell on the top ridge of the basin, it would flow to the bottom center and pool. If water fell on the sides of the basin, the water would also flow to the center and pool. All water falling within the basin will pool in the bottom center. Landscapes are composed of basins. The top of basins are ridges surrounding a low-lying area such as a wetland, pond, or lake where water will pool.

Using Topographic Maps to Delineate Watersheds

Suppose you needed to delineate the boundaries of the watershed surrounding the river network illustrated in the model below. The watercourse consists of a stream flowing from high to low elevation, and ending with a ponding area. The watercourse is depicted with a dotted line. To delineate a watershed, you need to mark the areas of highest elevation on both sides of the watercourse (indicated with stars on the figure to the right). Then, connect the points. Watershed boundaries cross contours at right angles because they will occur at the top of ridges. Refer to Chapter 5 to review topographic characteristics.

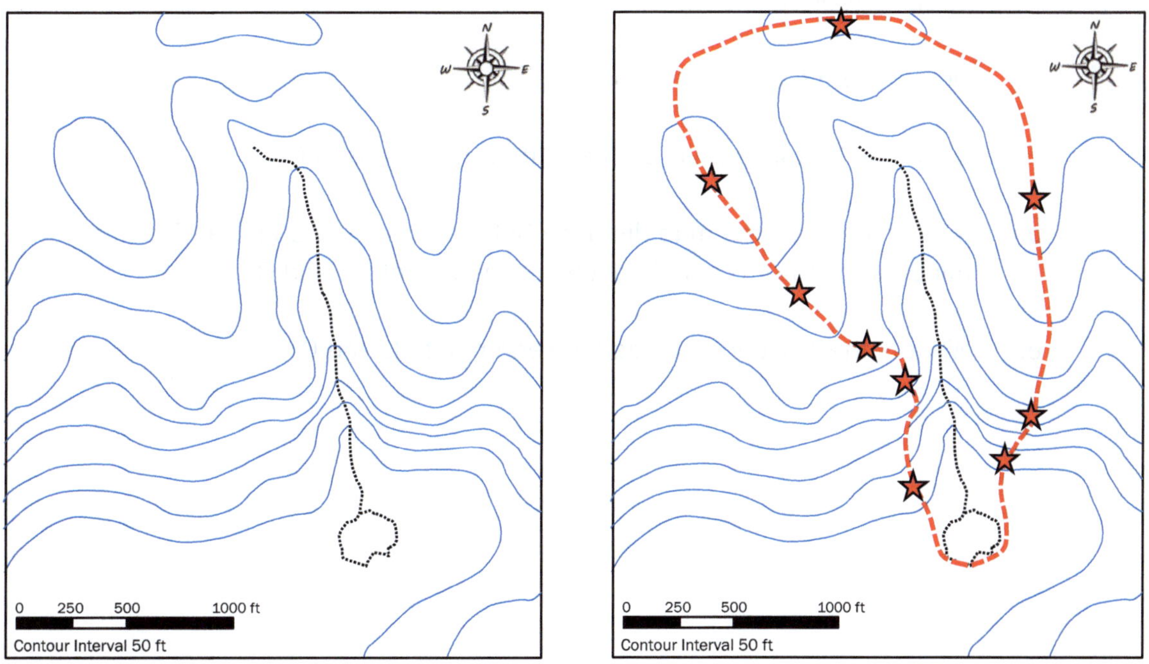

SUMMARY

Topographic maps are useful for understanding the shape and elevation of earth's features. Surface maps, such as contour maps, depict topographic features using lines of equal elevation. Contour maps may be created using elevation data collected systematically in a grid, where the distance between grid cells defines the map resolution. Such maps may be used to understand and delineate boundaries of watersheds, which are basins of land where surface water would drain into the same place (i.e., the bottom of the basin).

TEST YOUR SKILLS

1. In the Contour Delineation map, delineate 1-ft contours within the target area. Each number represents the elevation in feet. The shaded area at 91 ft represents a pond.

2. Use the Watershed Delineation map to do the following (note that elevation is in meters):

    a. Delineate the watershed boundary around two ponds and their associated streams (dotted lines).

    b. Determine the elevation of point A.

    c. Describe the topography on the path from point B to C.

    d. If you measured the map distance between points B and C as 7 cm, determine the on-the-ground distance between points B and C.

    e. What is the slope in degrees between points B and C?

Contour Delineation

Distance east (ft)	Distance north (ft)														
	0	20	40	60	80	100	120	140	160	180	200	220	240	260	280
0	99	99	97	96	96	96	96	96	96	96	97	98	98	99	100
10	98	98	96	96	95	96	96	96	96	95	96	97	98	99	100
20	97	98	96	96	95	96	96	96	96	96	96	97	97	98	99
30	97	97	96	95	95	95	96	96	96	96	96	97	98	98	98
40	96	96	95	95	95	95	96	96	96	96	96	97	98	98	98
50	96	95	95	95	95	95	95	95	96	96	96	97	97	97	98
60	95	95	96	95	95	95	95	95	96	95	96	96	97	97	97
70	95	94	94	94	94	94	94	94	94	94	95	96	96	97	97
80	95	94	93	93	93	93	93	93	93	93	93	95	96	96	97
90	94	93	93	93	93	93	93	93	93	93	93	94	95	96	96
100	95	95	93	93	93	93	93	93	93	93	93	93	94	95	96
110	95	93	92	92	92	92	92	92	92	91	91	92	93	95	96
120	96	92	91	91	92	91	91	91	91	91	91	91	91	93	96
130	96	91	91	91	91	91	91	91	91	91	91	91	91	93	95
140	96	91	91	91	91	91	91	91	91	91	91	91	91	93	95
150	96	91	91	91	91	91	91	91	91	91	91	91	91	92	94
160	96	93	91	91	91	91	91	91	91	91	91	91	91	92	93
170	95	93	91	91	91	91	91	91	91	91	91	91	91	92	92
180	95	93	91	91	91	91	91	91	91	91	91	91	91	92	92
190	95	93	91	91	91	91	91	91	91	91	91	91	91	91	92
210	93	93	92	91	91	91	91	91	91	91	91	91	91	91	92
220	93	92	92	92	91	91	91	91	91	91	91	91	91	92	92
230	92	92	92	92	91	91	91	91	91	91	91	91	91	92	93
240	92	92	92	92	91	91	91	91	91	91	91	91	91	92	93
250	92	92	92	92	92	92	92	92	92	92	92	92	92	93	94

Watershed Delineation

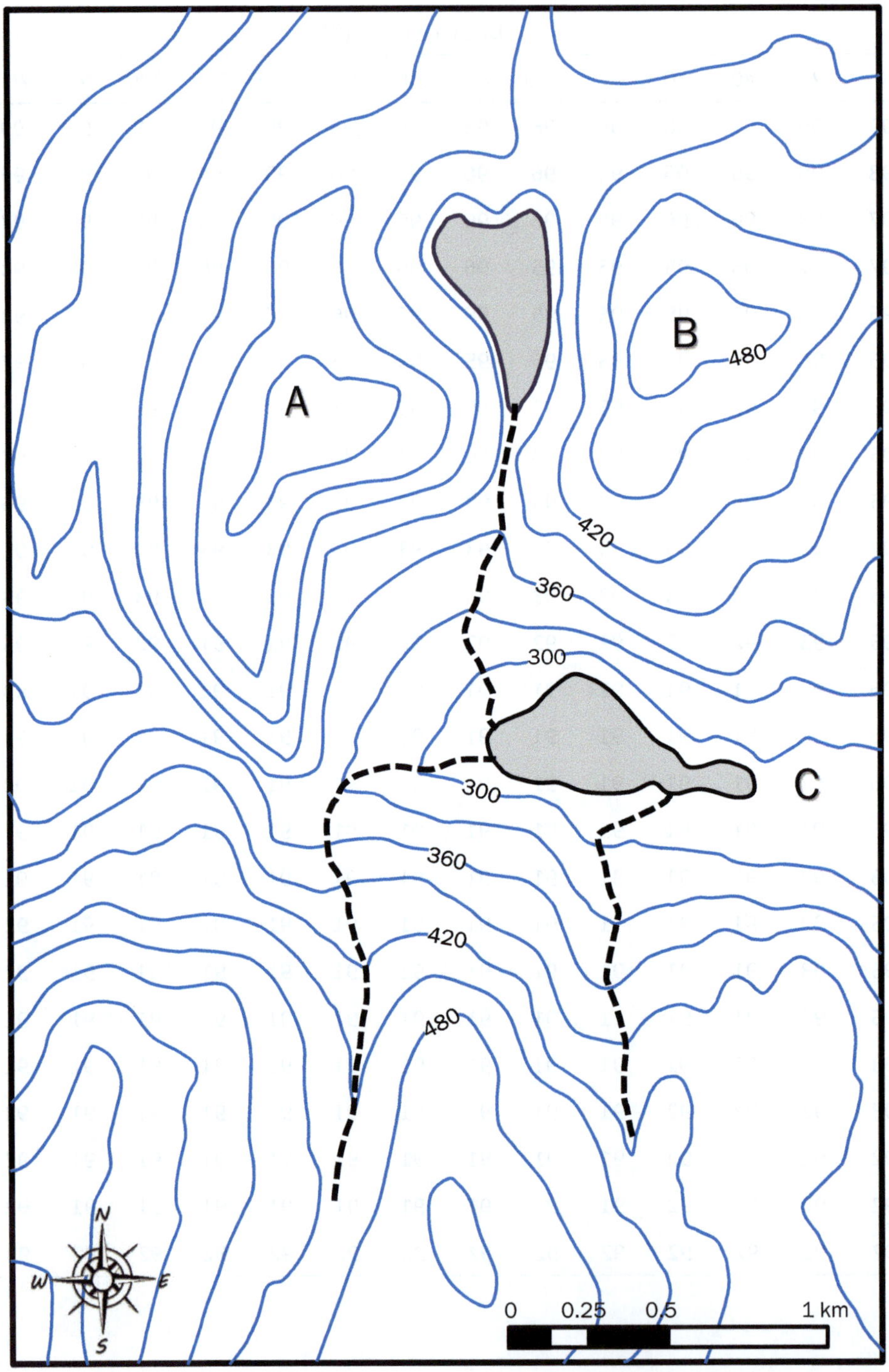

REFERENCE

Brooks, K. (2012). *Hydrology and Management of Watersheds*, 4th Edition. Des Moines, IA: Iowa State University Press.

9

Using Math for Solving Natural Resources-Related Problems

Problem solving is an essential skill to have in natural resources professions. Oftentimes, you will need to use mathematical principles to solve problems so that you have quantitative and defensible answers. For instance, if you were a forester and constantly overestimated tree volume, how would you explain to your client that it's your fault they didn't get as much money as you predicted from timber harvesting? If you were an environmental scientist and were in charge of protecting a stream from continuous sedimentation due to erosion on the banks, how would you explain to your boss if you overestimated the amount of erosion-control materials to purchase and you bought far too much?

In essence, precision in problem solving saves time, money, and credibility. This is where math comes into play.

LEARNING OBJECTIVES

1. Understand how to calculate area and perimeter of various basic shapes
2. Use the Pythagorean Theorem to solve problems associated with right triangles
3. Use trigonometry, including sine, cosine, and tangent (SOH CAH TOA), to solve problems associated with right triangles
4. Recognize appropriate mathematical methods for solving some natural resources–related problems

Use the guide below to help you determine which mathematical principles you should use to solve problems and help you practice your problem-solving skills with scenarios related to natural resources management.

NOTE: TO PROCESS ANGLES IN UNITS OF DEGREES, YOU MUST MAKE SURE YOUR CALCULATOR IS SET TO DEGREES AND NOT RADIANS.

Review of Basic Geometry

Perimeter

Rectangle	$Perimeter = 2 * length + 2 * width$
Circle	$Circumference = 2\pi * radius$ $Circumference = \pi * diameter$
Triangle	$Perimeter = length + length + length$

Area

Rectangle	$Area = length * width$
Circle	$Area = \pi * radius^2$
Triangle	$Area = \frac{1}{2} * base * height$

Angles

Rectangle	Angles of a rectangle add to 360°
Circle	Angles of a circle add to 360°
Triangle	Angles of a triangle add to 180°

Right triangles: Finding length of sides

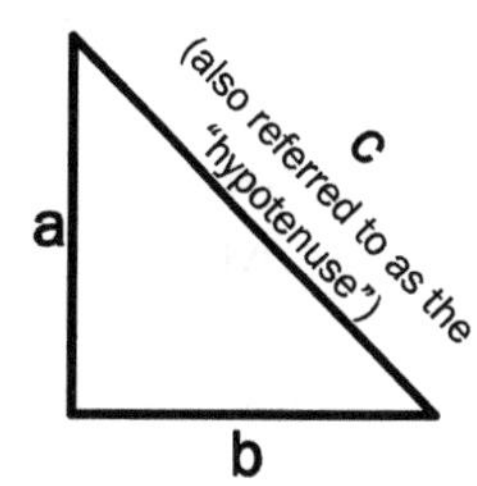

Pythagorean Theorem $a^2 + b^2 = c^2$

Where,
a = length of side, b = length of side, c = length of hypotenuse

When to use it

$a^2 + b^2 = c^2$ when you know the lengths of two sides of a triangle
AND you need to find the length of the third side
AND when no angles are needed

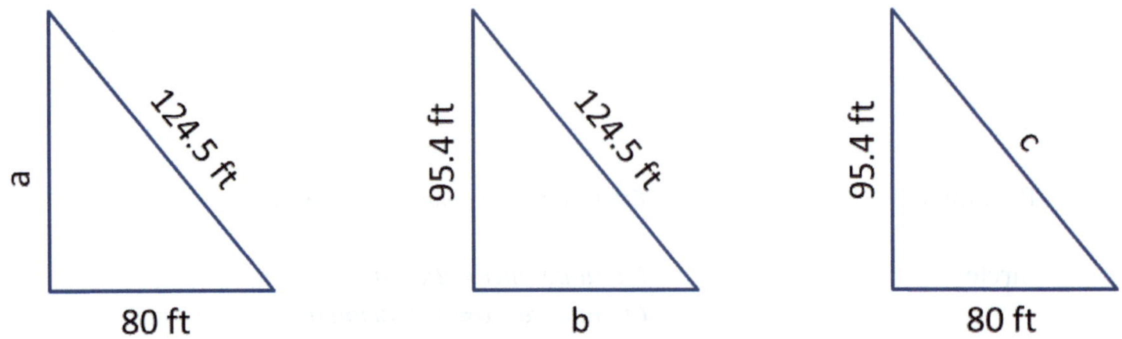

$$a^2 + 80^2 = 124.5^2 \qquad 95.4^2 + b^2 = 124.5^2 \qquad 95.4^2 + 80^2 = c^2$$

Review of Basic Trigonometry (Circles)

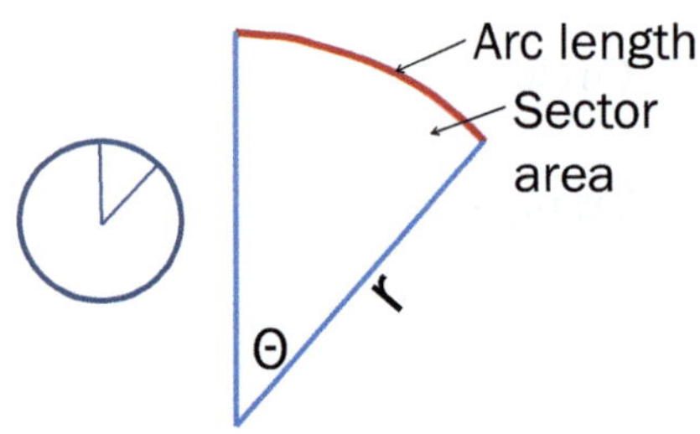

Arc length $$Arc\ length = \frac{\Theta}{180} * \pi * radius$$

Sector area $$Sector\ area = \frac{\Theta}{360} * \pi * radius^2$$

Review of Basic Trigonometry (SOH CAH TOA)

SOH The **sine** of an angle on a right triangle is equal to the length of the opposite side divided by the length of the hypotenuse.

Use when you know an angle and need to find the length of the opposite side of the angle or the length of the hypotenuse

OR when you are trying to find the angle associated with the opposite side of a triangle with known length

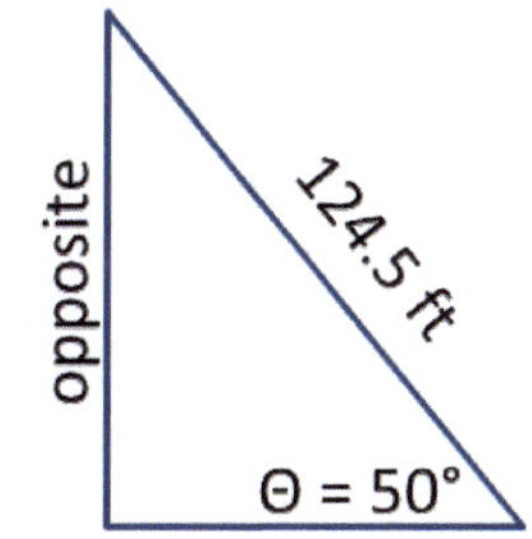

$$sin(50°) = \frac{opposite}{124.5}$$

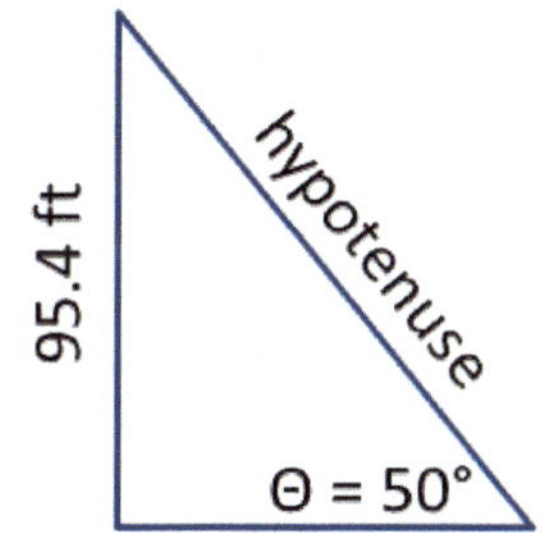

$$sin(50°) = \frac{95.4}{hypotenuse}$$

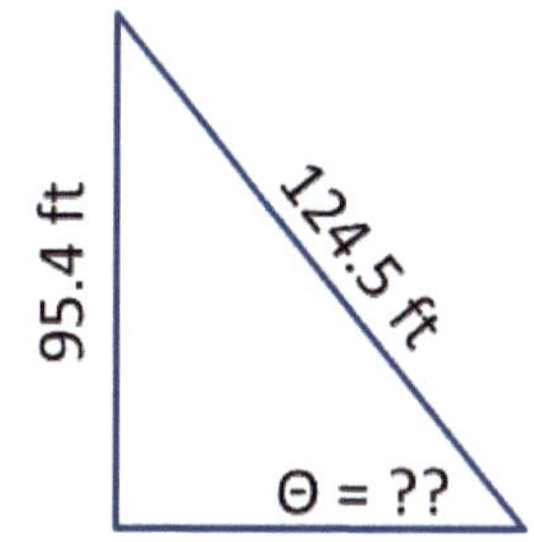

$$sin(\Theta) = \frac{95.4}{124.5}$$

CAH The **cosine** of an angle on a right triangle is equal to the length of the adjacent side divided by the length of the hypotenuse.

Use when you know an angle and need to find the length of the adjacent side of the angle or the length of the hypotenuse

OR when you are trying to find the angle associated with the adjacent side of a triangle with known length

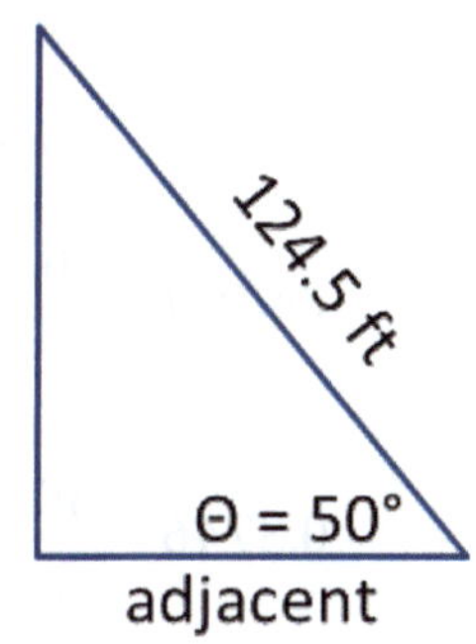

$$cos(50°) = \frac{adjacent}{124.5}$$

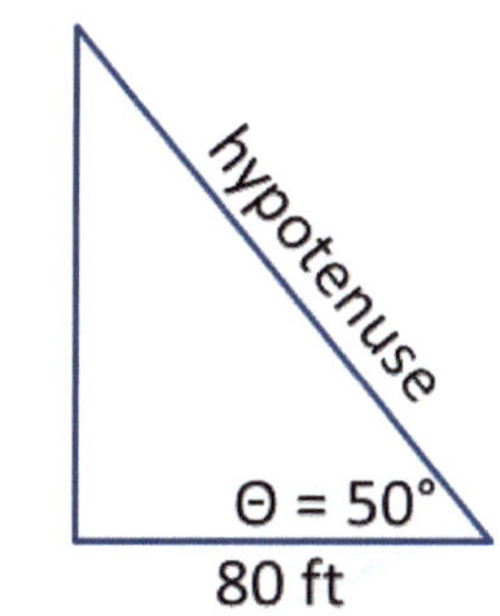

$$cos(50°) = \frac{80}{hypotenuse}$$

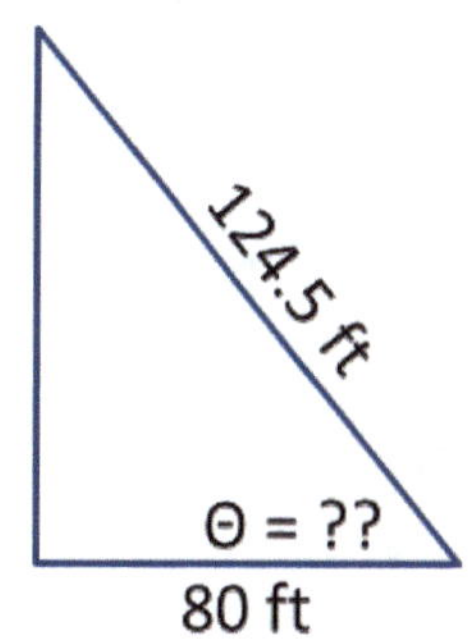

$$cos(\Theta) = \frac{80}{124.5}$$

TOA The **tangent** of an angle on a right triangle is equal to the length of the opposite side divided by the length of the adjacent side.

Use when you know an angle and need to find the length of the opposite side or the length of the adjacent side

OR when you are trying to find the angle associated with the opposite side of a triangle when the lengths of the adjacent and opposite sides are known

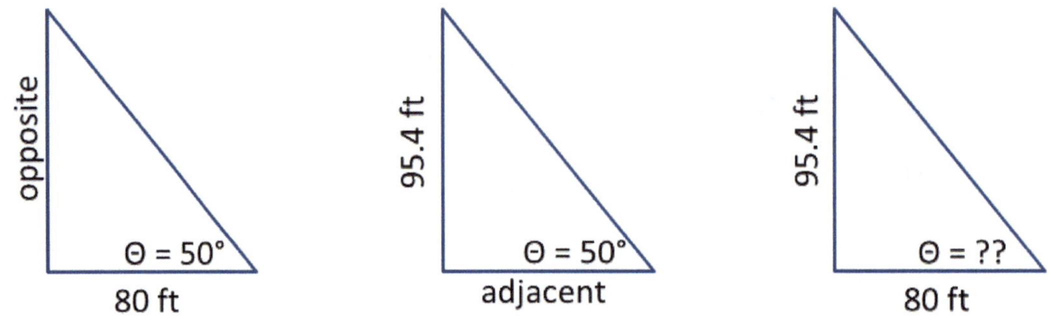

$$tan(50°) = \frac{opposite}{80} \qquad tan(50°) = \frac{95.4}{adjacent} \qquad tan(\Theta) = \frac{95.4}{80}$$

Practice Problem Demonstrations

Applications of geometry and trigonometry for solving problems in natural resources. First, some basic problems.

1. Solve for x

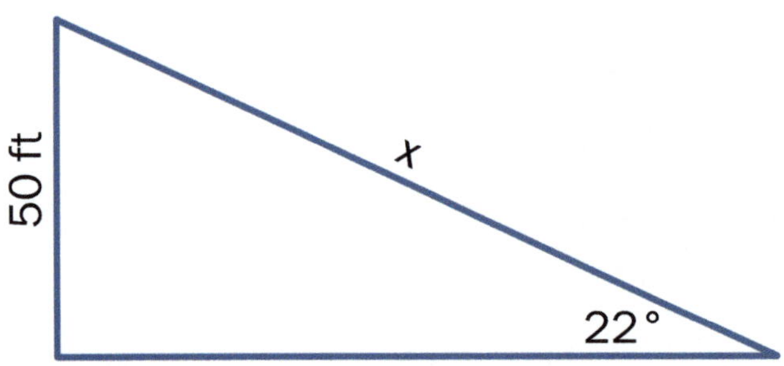

$$\sin(22) = \frac{50}{x}$$

$$0.3746 = \frac{50}{x}$$

$$0.3746x = 50$$
$$x = 133.5 \text{ ft}$$

Application: *You could use this type of calculation to find the on-the-ground distance if you have to traverse down a hill on which the landscape changed in elevation by 50 ft and the slope of the land was 22°.*

2. Solve for x

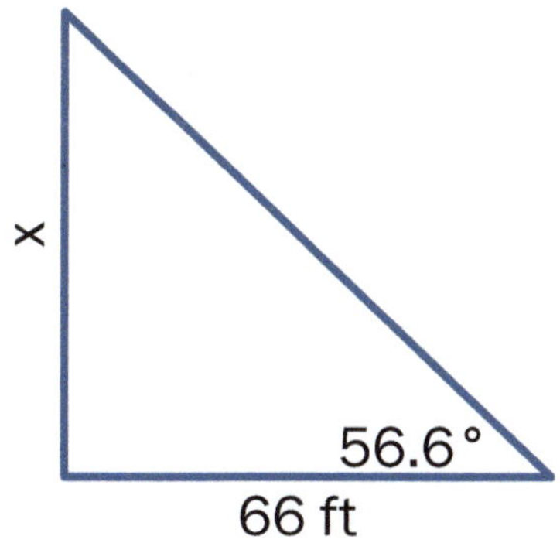

$$\tan(56.6) = \frac{x}{66}$$

$$1.5166 = \frac{x}{66}$$

x = 100 ft

Application: *You could use this type of calculation to find heights of objects if you knew the on-the-ground distances between where you were standing and the objects. For instance, common height measurements in natural resources fields are tree height, height of nests, and height of buildings. You can also use this type of calculation for distances (i.e., distance across a river or lake).*

3. Solve for x and y

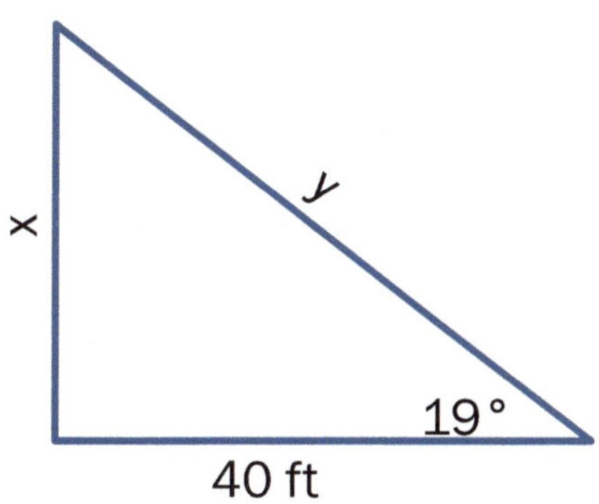

$$tan(19) = \frac{x}{40}$$

$$0.3443 = \frac{x}{40}$$

x = 13.8 ft

$$cos(19) = \frac{40}{y}$$

$$0.9455 = \frac{40}{y}$$

y = 42.3 ft

Application: *If you measured the flat distance between two points on a topographic map as equivalent to 40 feet on the ground, you could use the cosine function to find the on-the-ground distance between the top and bottom of the hill. You would need to know the slope of the hill, though. Also, if you knew the slope, you could calculate the elevation change between the top and bottom of the hill using tangent.*

4. Solve for θ

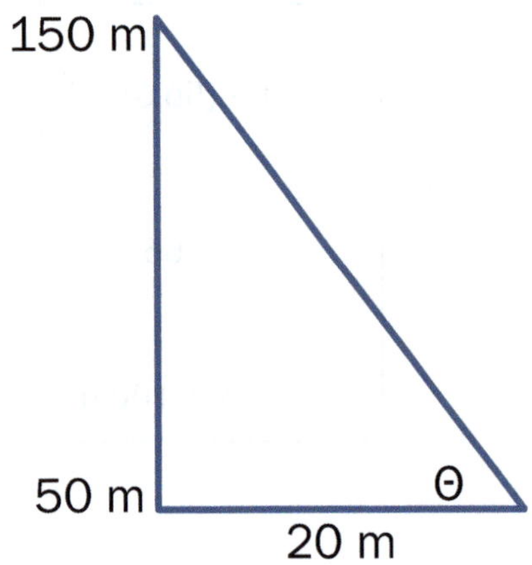

$$\tan(\theta) = \frac{100}{20}$$

$$\tan(\theta) = 5$$

$$\theta = \tan^{-1}(5)$$

$$\Theta = 78.7^\circ$$

Application: *If you had a topographic map, and you were looking at a mountain or hill of some sort, you could calculate the slope of that hill by determining the difference in elevation between the top and bottom of the hill and measuring the horizontal distance (i.e., the width of the base of the hill) between the top and bottom of the hill.*

5. What is the perimeter of the property boundary indicated below?

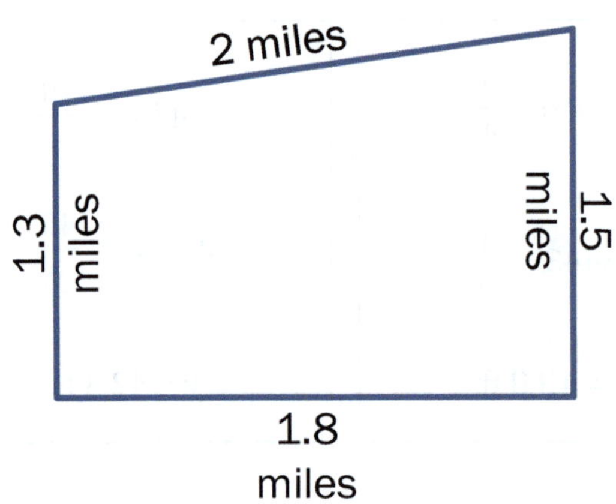

1.3 + 2 + 1.5 + 1.8 = 6.6 miles

Application: *Measuring the perimeter of forested areas, property boundaries, fields, etc. Determining the linear length of fencerows, edges, road networks, etc.*

6. If the square area bounded by roads in the figure below is 1 section, what is the perimeter of the property boundary labeled A indicated below (in miles)?

Source: U.S. Department of Agriculture.

North boundary = 0.125 mile
East boundary = 0.5 mile

Perimeter = 2(0.125) + 2 (0.5)
Perimeter = 1.25 miles

Application: *Determining areas and distances on the ground from aerial images.*

7. In the same image, what is the area of the property boundary labeled B indicated above (in hectares)? What is the east–west distance across B in km?

Parcel B is 80 ac
80 ac * 0.405 = 32.4 ha

E–W distance is 0.5 mile
0.5 mi * 1.6 = 0.8 km

Application: *Determining areas and distances on the ground from aerial images.*

8. What is the area (in ac) of the land parcel indicated below if the scale is 1:3,540?

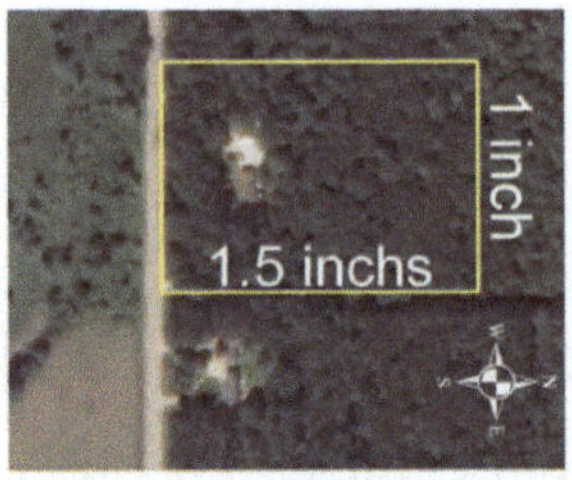

1 inch on the map = 3,540 inches on the ground (442.5 ft)
1.5 inches on the map = 5,310 inches on the ground (295 ft)

Therefore, the area is 442.5 * 295 = 130,537.5 ft^2
130,537.5/43,560 = 3 acres

Source: U.S. Department of Agriculture.

Application: *Determining areas and distances on the ground from aerial images using scale.*

9. If your destination was due south (180°), but you forgot to account for a 10° East declination, how far (in miles) would you be off course if you walked two miles?

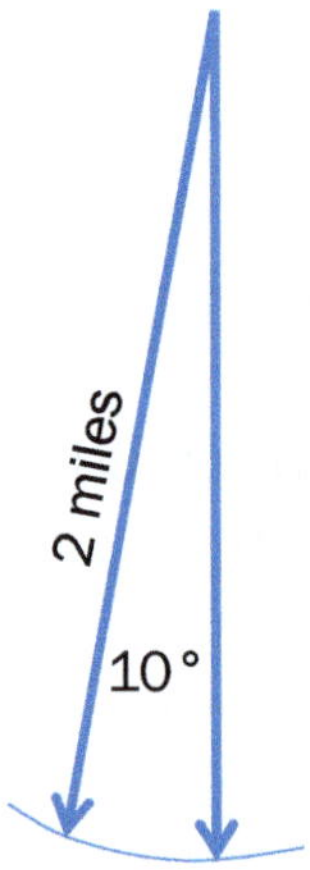

$$\text{Arc length} = \frac{\theta}{180} * \pi * radius$$

$$\text{Arc length} = \frac{10^\circ}{180} * \pi * 2\ miles$$

$$\text{Arc length} = 0.35\ miles$$

SUMMARY

Math, including algebra, geometry, and basic trigonometry, is frequently used to solve problems in natural resources management. Right triangles may be used to model some of these problems; for example, tree height, slope, and distance are some instances where right-triangle math may be applied. The Pythagorean Theorem, or principles of trigonometry including sine, cosine, and tangent, may also be used to solve such problems. Recognizing and using appropriate mathematical principles is important in providing quantitative assessments of many natural resources-related scenarios.

TEST YOUR SKILLS

Math Applications to Natural Resources Problem Solving

Please solve the following problems.

1. You are trying to get onto a roof. You know the distance from the ground to the roof is 15 ft, and the base of any ladder you put up must be at least 3 ft from the wall. What is the minimum length the ladder must be to reach the roof?

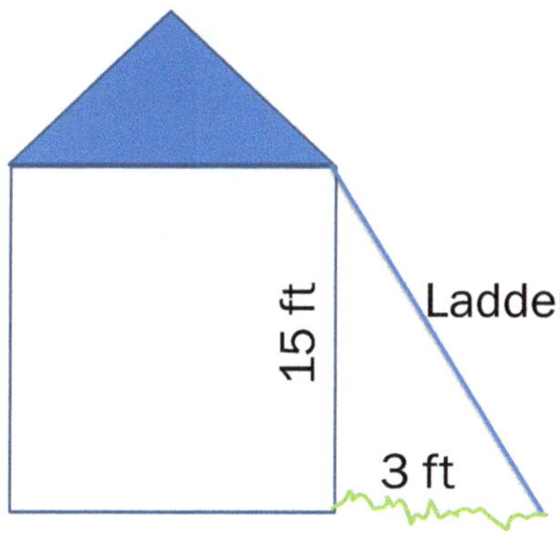

2. You are estimating the height of trees. You move back from the base of each tree by two chains. You measure the angle needed to see the top of the tree using an instrument (clinometer) that measures angles. Assume you are on flat ground and the angle you measure is 36 degrees. What is the height of the tree?

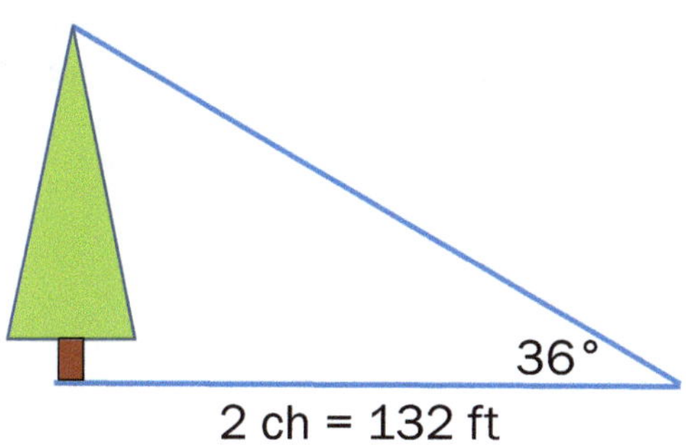

3. You are planning a photo flight. You have an area of two sections (1,280 acres) to cover. Given the altitude at which you will fly and the camera system you will use, you will cover 65,340 ft^2 with each photo taken. What is the minimum number of photos you will need to take to cover the entire area?

4. If you measure 223 mm on a map that represents the distance of a road that you know is 1.2 miles long, what is the number of inches (on the photo) per mile (observed on the ground)?

5. You are a wildlife biologist charged with net-gunning ducks so you can age, sex, and band them. You have duck decoys set up in shallow water, and have cracked corn sprinkled on the shore. Before you start calling ducks, you need to determine how far from the center of your decoy spread to set the cannon. Your cannon can shoot a net 70 feet, but you want to aim about 15 feet over the water to make sure you capture all the ducks as they attempt to escape. At what angle do you need to set your cannon? How far from the center of the decoy spread (in feet) does the cannon need to be?

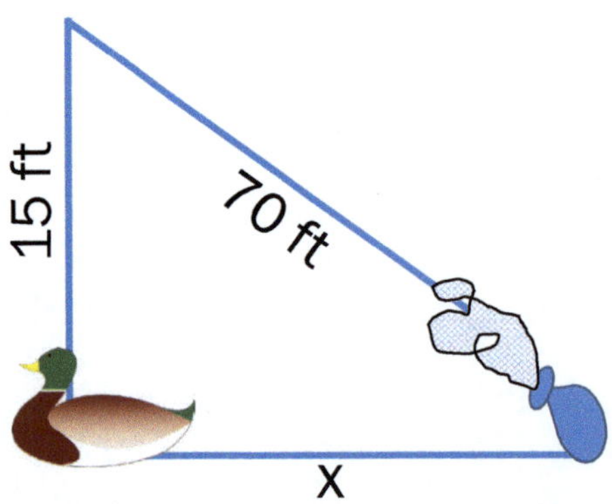

Source: http://pixabay.com/en/mallard-duck-bird-waterfowl-swim-583365/.

6. You work for a water quality company that helps minimize erosion problems caused by construction. You are investigating the site for a new road and observe that there is a 0.75-mile stretch next to a slope that will need erosion control. Surveyors have measured the elevation at the top of the slope as 640 feet above sea level. Within a horizontal distance of 40 feet, the elevation decreases to 590 feet, where it flattens out at the bottom. To control erosion, you recommend using a soil-stabilizing material over the entire slope. How much stabilizing material do you need (in square feet) to cover the entire slope? What is the slope (in degrees)?

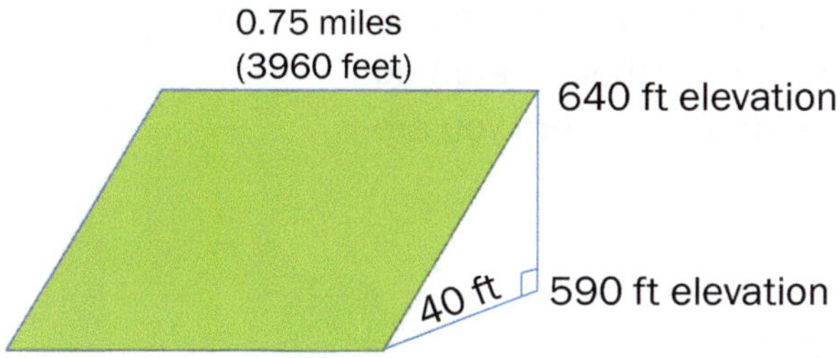

7. A mountain goat is standing on a mountain with a slope of 61°. You are standing at the bottom of the mountain. You estimate with a powerful rangefinder that the goat is 800 ft up the slope from you. How high (in feet) from the base of the mountain is the goat?

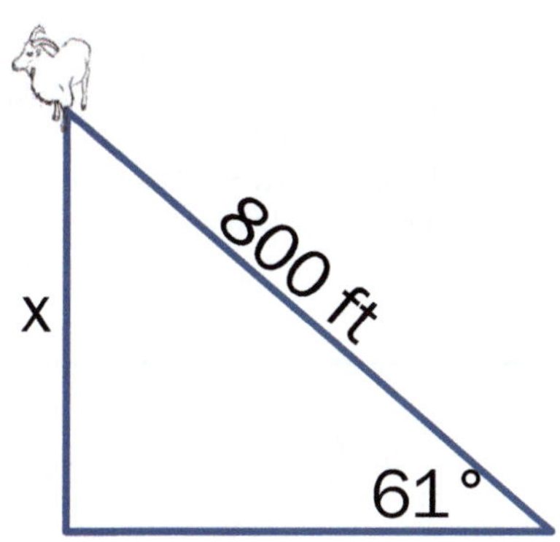

Source: http://pixabay.com/en/goats-kid-animals-mountain-goat-713425/.

8. Your research buddy, Poindexter, measured a distance and azimuth off a map from the field station where you worked to a spot along a stream where you were going to sample aquatic invertebrates. Poindexter told you the azimuth was 165°, and the distance to the stream was 3 miles. You paced the specified distance along the specified azimuth, but did not come to a stream. Just then, you realize Poindexter did not account for declination. In your location, declination is 6° west. How far from the stream are you (in feet)? In what general direction (north, northeast, etc.) do you need to travel to get to the stream from where you are currently standing?

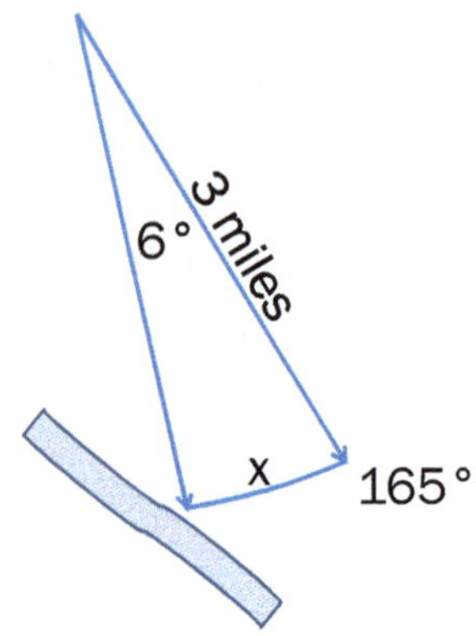

9. You are an ornithologist and want to record the nesting success of woodpeckers. In your study area, you measure the height of a tree cavity where you know the birds are nesting as 38 feet. To take a good picture of the nest at the bottom of the cavity, the camera must be positioned at an angle. If you have a camera mounted to the top of a pole that can extend 50 feet, how far back from the base of the tree must you stand to position the camera if the pole is extended all the way?

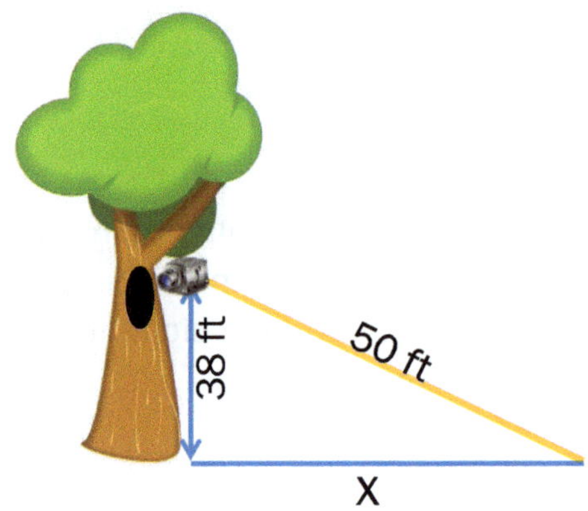

Source: http://pixabay.com/en/tree-environment-ecology-nature-146748/.
Source: http://pixabay.com/en/camera-equipment-lens-photography-151024/.

10. On a map, you measure 1 inch between point A and point B. The map scale is 1:2,500. There is a 90-ft elevation difference between A and B. What is the on-the-ground distance between A and B?

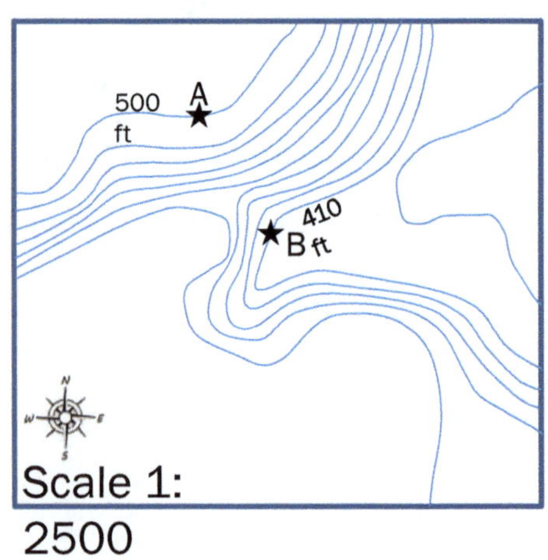

11. You work for the EPA and were charged with assessing the pollution impact of a chemical spill from a truck that overturned on the highway. You measure levels at the spill and at several distances away from the spill. You found that there was no trace of the chemical at distances exceeding 0.2 kilometers from the spill. How large was the area impacted (in hectares)?

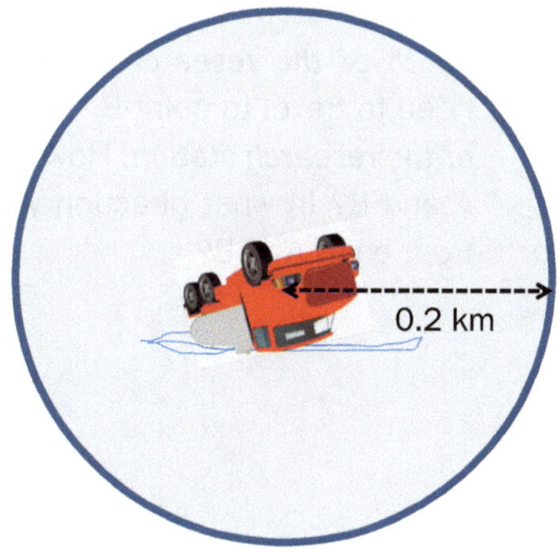

Source: http://pixabay.com/en/truck-lorry-vehicle-transport-303375/.

12. Using the diagram to the right, what is the area of Mr. Pace's land (in acres)?

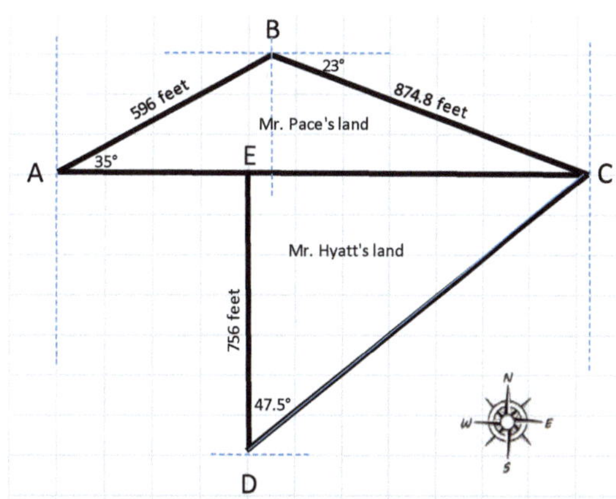

13. What is the perimeter of Mr. Hyatt's land (in meters)?

14. You are lost someplace. You know how to determine your latitude and longitude from the sun, but you are not sure when it is local noon. So, you find a straight stick and push it into the ground so it sticks up 2 meters. Then, you mark on the ground where the shadow falls throughout the day. The shortest shadow was cast at local noon, but you need the angle of the sun to calculate latitude. If the shortest shadow is 22 cm, what is the angle of the sun? If it is July 3, at what latitude are you? (Hint: use an analemma)

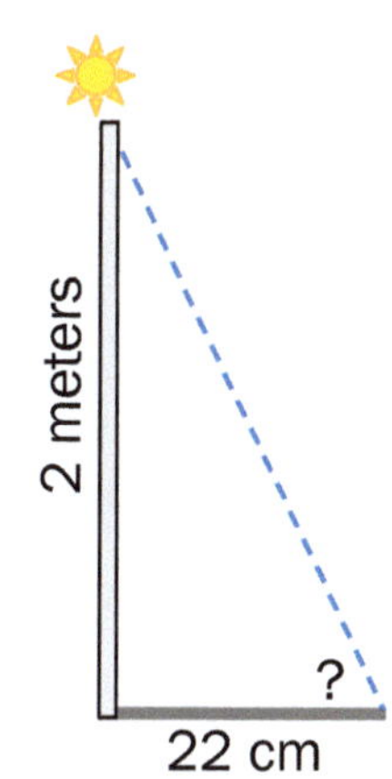

15. You are doing field work and need to travel from the research station to point A, which is 1.2 km ***due South of the research station***. From point A, you need to travel to point B, which is 1.3 km ***due West of the research station***. How far is it between points A and B? In what direction would you travel to get from point A to B?

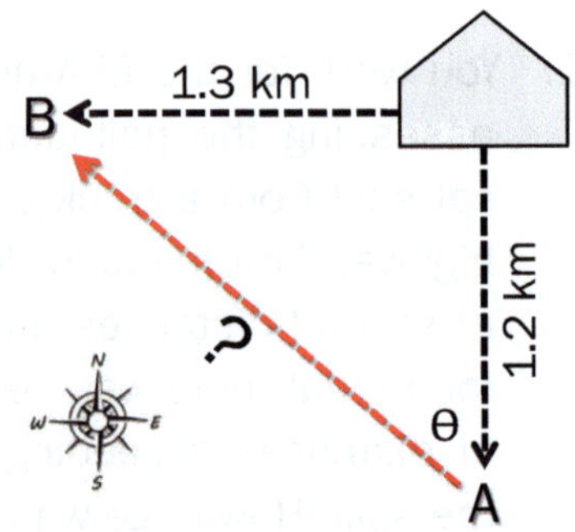

10

Remote Sensing and Aerial Photos

By now, you should be well aware that taking measurements from maps and being able to identify spatial relationships from maps are critical for being a good professional in a natural resources–related field. Many maps are created using data acquired from satellites or other aerial platforms via remote sensing. **Remote sensing** is the acquisition of data from a distance. There are many types of remote sensing, and we use it every day. X-rays, mobile phones, digital cameras, remote starters on vehicles, and global positioning systems are a few examples of technology using remote sensing.

Understanding aerial photos and imagery captured using remote-sensing techniques is critical in developing your professional skills. Aerial photos and imagery are used to study land changes over time, depict details about the earth's surface, create maps, etc. There is a difference between **aerial photos** and **aerial imagery**. Aerial photos are hard-copy pictures, whereas aerial imagery is digital. With advances in technology such as use of kites or unmanned aerial vehicles (drones) for capturing imagery, and data-sharing capabilities, aerial photos and imagery are largely available and much of it is free. Thus, it is likely that you will come across aerial photos and imagery in the fields of wildlife management, forestry, environmental science, urban and landscape planning, fisheries, watershed management, and many others. When you do, you will need to understand how to interpret them. In this chapter, you will gain experience using aerial photos and identifying specific information from them. An aerial photo is no longer just a cool picture—it contains DATA! Valuable data! And you can glean much information from aerials if you understand clearly how to interpret them and take measurements from them.

Photogrammetry is the science of making scale-accurate measurements from photos for the primary purpose of mapmaking.

LEARNING OBJECTIVES

1. Understand the process of remote sensing and how it is used to capture aerial photos/imagery
2. Take horizontal measurements from aerial photos and interpret them

3. Measure the height of objects from aerial photos
4. Convert photo distance to earth distance using scale

What Is Remote Sensing?

Remote sensing is collection of data from a distance. In terms of acquisition of aerial data, remote sensing refers to earth observation from satellites or aircraft. Data from aerial imagery are interpreted by natural resources professionals to determine the nature of features on the earth and their significance. For example, imagery may be used to determine the number, density, and spatial arrangement of features within an area. Imagery may also be used to classify landscapes according to meaningful categories such as land ownership or land-cover type.

Aerial imagery relies on the properties of light or near infrared energy within the **electromagnetic spectrum**. The electromagnetic spectrum is the range of all possible frequencies of electromagnetic radiation from low-frequency radio waves to very high-frequency gamma and cosmic rays. Ultraviolet light, visible light, and infrared energy represent the portion of the electromagnetic spectrum important in remote sensing.

Electromagnetic energy, including light or infrared wavelengths, is reflected, transmitted, absorbed, or refracted. **Reflected energy** bounces off objects. For example, objects appear to be the specific color that they most reflect. **Transmittance** is the fraction of energy that is passed through something. More light energy is transmitted through a clear glass window than a tinted opaque one. **Absorbed energy** is converted to something else. For instance, plants convert light energy to food through photosynthesis. **Refraction** refers to the bending of wavelengths due to a change in velocity as it passes through different media. For instance, as light passing through water, it slows and bends.

There are two primary remote sensing systems: passive and active. **Passive systems** rely on light energy from the sun. For instance, satellite imagery is an example of data acquired through a passive system. Alternatively, **active systems** generate an energy signal and detect the energy returned. Doppler radar, for instance, is an example of an active remote sensing system to understand weather conditions.

Types, Acquisition, and Geometry of Photos

Aerial photos can be classified into two types: vertical and oblique. In **vertical photos** (below left), the camera is perpendicular to the ground; the camera captures the photo of the landscape from directly overhead. In **oblique photos**, the camera is not perpendicular to the ground; thus the picture of the landscape is at an angle. In low oblique photos, the horizon is not visible. In high oblique photos (below right), the horizon is visible.

There are several differences between vertical and oblique photos. First, scale differs between vertical and oblique photos. In vertical photos, the scale is essentially constant, whereas the scale varies across oblique photos. Notice in the oblique photo that the objects toward the bottom of the photo are closer (larger) than those toward the top of the photo. Thus, the scale at the bottom of an oblique photo is larger than the scale toward the top. Because scale is constant in vertical photos, it sometimes may be used as a map if a grid system (i.e., coordinates) is added. Second, objects may be more easily recognized in oblique photos than in vertical photos. In vertical photos, the camera is directly perpendicular to the ground. Thus, the photo depicts the top of objects, whereas in oblique photos, the sides of objects may be visible, and may aid in recognition and identification of objects. Third, some objects such as caves may be visible only in oblique photos, because they may be concealed from above in vertical photos. Fourth, it is easier and more accurate to measure direction in vertical photos than it is in oblique photos. For the remainder of this chapter, we will be working with measurements on vertical photos.

Before we begin taking measurements on vertical photos, it is important to understand their geometry and how they are acquired. The geometric center of the photo is called the **principal point**. The point directly beneath the camera is called the **nadir**. Halfway between the principal point and the nadir is the **isocenter**. In a true vertical photo (i.e., where the camera is perfectly vertical and not tilted at all from wind turbulence or drift), the principal point and the nadir are at the same location. For simplicity to demonstrate measurements from photos, we will assume that the principal point and nadir are in the same location on the photos we use in this chapter. **Fiducial marks** on the photo are indicated in the corners and halfway between. Flight information, date, and sequence are also indicated on the photo.

Prior to photo acquisition within a target area, a flight plan is made. This plan includes starting and ending points of a flight path, the interval at which photos are taken, and determination of the amount of overlap on successive photos. Successive photos commonly have 60% overlap between them. Overlap is necessary for stereoscopy. **Stereoscopy** is a process for creating three-dimensional depth by viewing an object in two photos taken from different vantage points. For example, the only visible part of an object directly below the camera (i.e., the nadir) is the top. Imagine if you were hovering directly over a tree. You would be able to see the top of the tree, but not the sides or bottom. Radiating outward from the nadir, sides of objects are visible. Thus, toward the edges of photos, displacement occurs. **Displacement** is a shift in the position of objects on aerial photos or aerial images due to the perspective from the camera. Displacement is evident when the top and bottom of the same object are not in the same location because the camera's position was not directly above that object. Displacement is the reason we can take vertical measurements such as height from a photo. In the photo of Ground Zero above, notice that the camera was directly above the point in the center of the photo. There is no displacement there. However, radiating outward from the nadir, there is more and more displacement. The sides of buildings are evident and the top and bottom of each building is not in the same location; that is, there is a shift in position between the top and bottom of the same structure. Taking measurements of the amount of displacement on an object or viewing the same object from two different stereoscopic photos allows us to take vertical measurements.

Source: http://commons.wikimedia.org/wiki/File:Wtc-photo.jpg.

Horizontal Measurements from Aerial Photos

In previous chapters, you used scale to calculate distances between objects on maps. Scale is essential for estimating from maps or photos horizontal earth distances. Recall that scale represents the relationship between map or photo distance and earth distance. Several techniques may be used for calculating horizontal distances or area, but understanding scale is essential for each. For instance, distance may be computed directly from the scale, or by using a ratio between the photo plus known earth distance of a feature and the photo plus unknown earth distance of another feature. Let's practice using scale to compute horizontal distances directly.

Horizontal distance using scale

1. Suppose you were evaluating the growth of trees in an orchard to determine the effectiveness of an organic fertilizer. You used an unmanned aerial vehicle (UAV; aka drone) to capture vertical photos of the orchard at a scale of 1:175. If you measured the width of a tree to be 1.6 cm on the photo, how wide is the canopy of the tree in meters?

Source: http://pixabay.com/en/mallorca-field-landscape-191967/.

► To solve this problem, you need to set up a ratio using the photo scale, solve for the canopy width in centimeters, and then convert to meters. First, understand the structure of the ratio. According to the scale, 1 cm on the photo is equivalent to 175 cm on the earth. Thus,

$$\frac{1\ cm\ on\ the\ photo}{175\ cm\ on\ the\ earth} = \frac{1.6\ cm\ on\ the\ photo}{x\ cm\ on\ the\ earth}$$

Solving for the x, which is the unknown value of the tree canopy width, yields 280 cm. But the problem asked to solve for width in units of meters. So, a conversion is necessary (review Chapter 2). You know that there are 100 cm in 1 meter. Thus, **the tree canopy is 2.8 meters wide.**

Horizontal distance using ratios

2. Suppose you are working from a photo where the scale is unknown and you need to find the distance of Osborne Road, which measures 6 inches on a photo. You know that Warner Road is 2 miles long, and it measures 1 inch. How long is Osborne Road in miles?

► To solve this problem, simply set up a ratio with the photo and earth distance of Warner Road (known) equivalent to the photo and earth distance of Osborne Road (unknown).

$$\frac{1 \text{ in on the photo}}{2 \text{ miles on the earth}} = \frac{6 \text{ in on the photo}}{x \text{ miles on the earth}}$$

Solving for x, **Osborne Road is 12 miles long.**

Calculating area using scale

► Area is another horizontal measurement that may be computed from aerial photos. Once again, scale is necessary to calculate area. For example,

3. If a parcel of land on a map is 1 cm x 1 cm and the map scale is 1:50,000, how many hectares is it?

   a. First, find the area of this parcel of land on the photo. If it measures 1 cm x 1 cm, the area is 1 cm^2.

   b. Next, find the area on the earth using the given scale. According to the scale, 1 cm on the photo is equivalent to 50,000 cm on the earth. It might be more intuitive to convert 50,000 cm to meters. Recall that ultimately, we want area in hectares. There are 100 cm per 1 meter. Thus, 50,000 cm equals 500 m. So the area on the earth is 500 m x 500 m, or 250,000 m^2.

   c. The final step is to convert square meters to hectares. Recall that there are 10,000 m^2 per 1 ha. Thus, **the area of this land parcel is 25 ha.**

$$\frac{250{,}000 \not{m}^2}{1} * \frac{1\ ha}{10{,}000 \not{m}^2} = 25\ ha$$

Calculating area using a dot grid

► Occasionally, it may be necessary to measure lengths of winding rivers, or area of parcels with irregular or curvy boundaries. In such cases, it is not feasible to measure length using a ruler. Other tools such as a **map scaler** or **planimeter** may be used to compute horizontal distances or areas of irregular shapes. These tools allow the user to trace features, and then use scale and mathematical principles to compute distance or area. Alternatively, a **dot grid** overlaid on a photo allows a quick estimation of area. Each dot falling within a boundary is equivalent to a specific area depending on scale. A summation of the total number of dots within the area multiplied by the area per dot allows determination of area. Advances in technology such as Geographic Information Systems have largely replaced use of map scalers, planimeters, and dot grids.

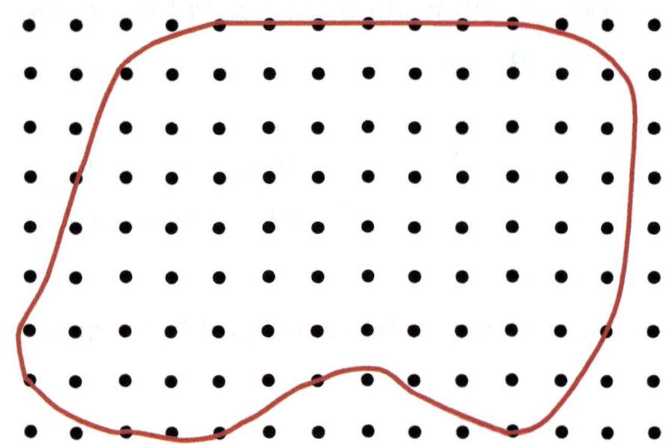

4. If each dot represents 0.65 acres, how large (in acres) is the lake indicated in the diagram?

   Note that to avoid underestimating or overestimating area, all dots within the diagram and *every other* dot directly on the boundary are included in the tally.

   *86 dots total * 0.65 acres* = **55.9 acres**

Vertical Measurements from Aerial Photos

Vertical measurements such as height of objects or elevation may be computed from aerial photos. Stereoscopic photos may reveal height when viewed through stereoscopes and measured with tools such as a parallax bar. Without stereoscopic photos, however, height may be calculated using the sun angle and shadow of objects, or by using measurements of displacement.

Sun-angle and shadow method

► If the angle of the sun at the time the photo was taken is known, height of objects can be determined by measuring the length of the object's shadow on the photo, using scale to convert shadow length in the photo to shadow length on the earth, and then using trigonometry to calculate height. This method is reliable on flat ground and for straight trees. If the ground was sloped or the trees were angled, height determined with the sun-angle and shadow method will be erroneous.

5. Suppose you were using the photo to the right to determine which trees were suitable in which to place nest boxes for a rare owl. The owl prefers to nest in trees that are at least 55 ft tall. The tree shadow measures 2 inches on the photo. Scale is 1:420, and the sun angle is 42°. Is the indicated tree suitable for a nest box?

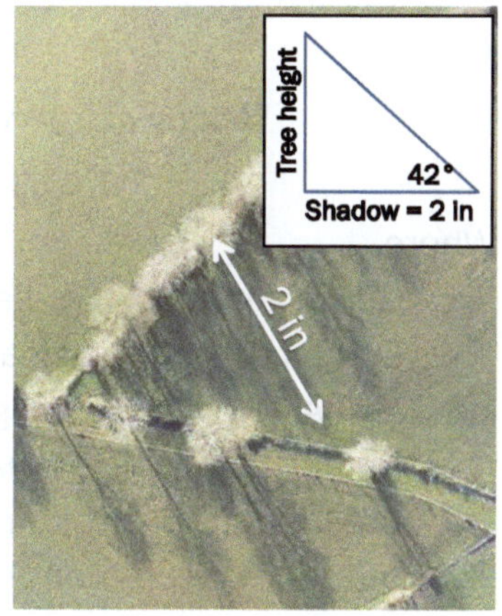

Copyright © Gordon Dunn (CC BY-SA 2.0) at https://commons.wikimedia.org/wiki/File:Country_Lane_-_geograph.org.uk_-_341131.jpg.

a. First, convert shadow length on the photo to shadow length on the ground using photo scale.

$$\frac{1\ inch\ on\ photo}{420\ inches\ on\ the\ earth} = \frac{2\ in\ shadow\ on\ the\ photo}{x\ in\ shadow\ on\ the\ earth}$$

The shadow length is 840 inches long on the earth, or 70 feet long.

b. Next, use trigonometry to calculate tree height. We know the sun angle and the length of the adjacent side. So, we can use tangent to solve for tree height.

$$\tan(\theta) = \frac{length\ of\ the\ opposite\ side}{length\ of\ the\ adjacent\ side}$$

$$\tan(42°) = \frac{tree\ height}{70\ ft}$$

$$\tan(42°)*70\ ft = tree\ height$$

$$tree\ height = 63\ ft$$

The tree is 63 ft tall. It is suitable as a nest box location for the owl.

Displacement method

► Height may also be determined by using measurements of displacement in the following equation:

$$Height\ of\ object = \frac{displacement * flying\ \ height}{radial\ distance\ from\ nadir\ to\ the\ top\ of\ object}$$

Where,

Displacement = the distance on the photo between the top and bottom of the object. Displacement is typically measured in millimeters or fractions of an inch.

Flying height = the height of the aircraft when the photo was taken.

Radial distance from nadir to the top of object = the distance on the photo between the nadir (the point directly beneath the camera) and the top of the object. The units of this measurement should be the same as that of displacement.

6. What is the height of the building in feet in the diagram to the right if displacement was 1.8 cm, the radial distance between the nadir and the top of the building was 5.5 cm, and the flying height of the aircraft was 100 ft?

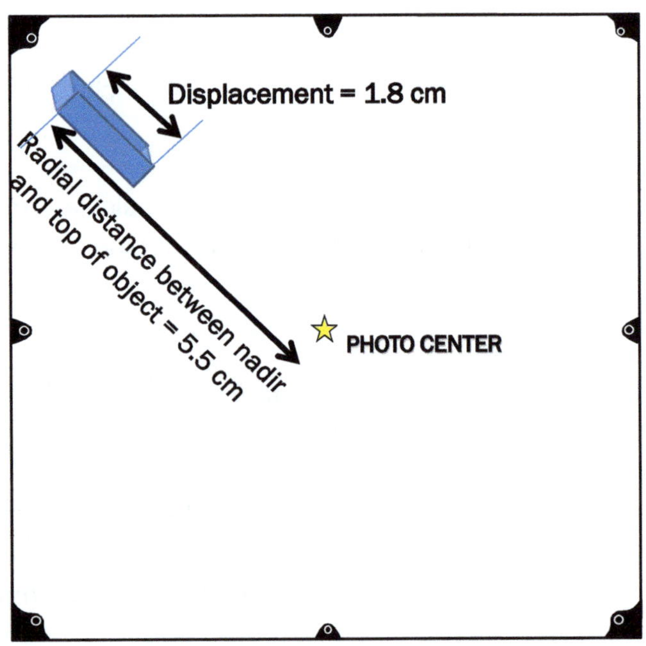

$$Height = \frac{displ * flying\ \ ht}{radial\ distance}$$

$$Height = \frac{1.8\ \cancel{cm} * 1{,}000\ ft}{5.5\ \cancel{cm}}$$

$$Height = 327\ ft$$

The building is approximately 327 ft tall.

SUMMARY

Remote sensing, or the acquisition of data from a distance, may be used to obtain aerial photos or images used to understand aspects of the earth. Aerial imagery relies on the properties of light or near infrared energy within the electromagnetic spectrum. Electromagnetic energy is reflected, transmitted, absorbed, or refracted, depending on what type of object or medium the energy is interacting with. Aerial photos or imagery capture the energy reflected off objects. Aerial photos and imagery may be vertical or oblique. Vertical photos are acquired when the camera lens is perpendicular to the earth's surface, whereas oblique photos are taken at an angle to the earth's surface. Understanding the geometric properties of aerial photos allows horizontal (i.e., distance and area) and vertical (i.e., height) measurements to be taken from them. Horizontal distances may be determined using scale and ratios. Vertical distances may be determined from the shadow lengths or displacement of objects.

TEST YOUR SKILLS

1. Suppose the shadow of the wind turbine in photo A measured 1.65 inches on the photo, the sun angle was 36°, and the photo scale was 1:2,000. Also suppose the shadow of the wind turbine in photo B measured 4.07 inches, the sun angle was 5°, and the photo scale was 1:5,900. Which wind turbine is taller? By how many feet?

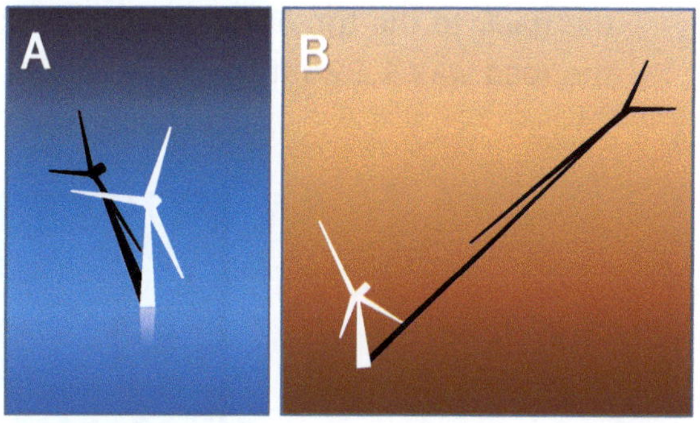

2. If the scale on the photo to the right is 1:150 and the width of the tree measures 2.1 cm on the photo, what is the actual width (in meters) of the tree canopy?

3. If a land parcel on a photo measures 0.5 inches across and is known to be 1,000 feet wide, how wide is a land parcel measuring 2.3 inches across?

4. A photo has a scale of 1:5080, the shadow of a water tower measures 1.4 cm, and the sun angle was 40.6 degrees, how tall is the water tower in feet? If the shadow from trees in a nearby forest measures 0.4 cm, how tall are the trees in feet?

5. You measure the distance between the top and bottom of a tree on a photo as 0.35 inches. The distance from the center of the photo to the top of the same tree is 3.2 inches. If the flying height was 600 feet, how tall is the tree?

6. How tall is the bridge tower (in meters) if you knew that the distance on the photo from the nadir to the base of the tower (yellow line) is 0.625 in, the distance from the nadir to the top of the tower (black line) is 0.969 in, and the flying height above the road was 1,325 ft?

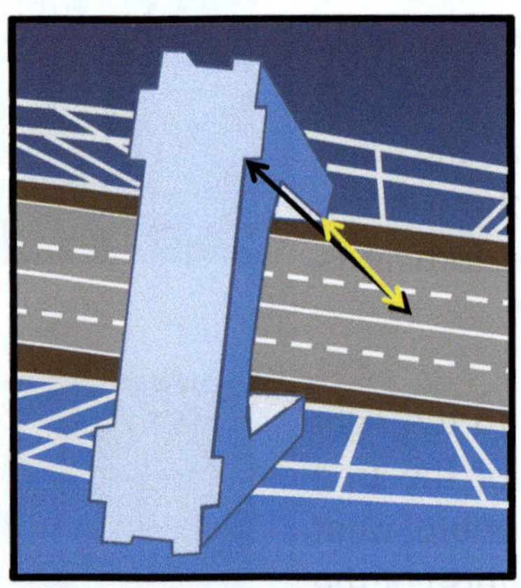

REFERENCE

Paine, D. P. & Kiser, J. D. (2003). *Aerial Photography and Image Interpretation*, 2nd Edition. Hoboken, NJ: John Wiley & Sons.

11

Land-Use/Land-Cover Classification

Photo or image interpretation is the process of identifying objects on aerial photos or imagery and determining their significance. Interpretation skills are necessary to classify land into either land-cover or land-use categories. Such classifications may be used in natural resources–management fields or earth science fields to understand current landscapes and changes over time. The classification of landscapes provides fundamental information about characteristics, activities, and statuses of earth's features.

Land cover refers to the physical or biological cover over the surface of the earth, including water, vegetation, bare soil, and/or components of the human-built environment (e.g., roads or structures). **Land use** generally refers to human activities such as agriculture, forestry, and building construction. While land cover may be observed directly in the field or by remote sensing, observations of land uses and their changes generally require the integration of natural and social scientific methods (e.g., expert knowledge or interviews with land managers) to determine which human activities are occurring in different parts of the landscape, even when land cover appears to be the same. For example, areas covered by mature woody vegetation may represent an undisturbed natural forest (e.g., use = wilderness), a forest reserve managed for endangered wildlife (e.g., use = conservation), regrowth following tree harvest (e.g., use = timberland), or a plantation of immature pecan trees (e.g., use = agriculture).

Land-use and land-cover (**LULC**) is a term used broadly to classify land into different categories describing either land cover or land use. These classifications are used widely by resource managers, social scientists, policymakers, law enforcement officials, and many others. There are many LULC Classification Systems in use today. Being familiar with the major ones will help you if you need to choose the most appropriate system for a project or job. Two major standardized national LULC classification systems are the Anderson Classification System (http://landcover.usgs.gov/pdf/anderson.pdf), which uses remotely sensed aerial photography to classify landscapes, and a classification system modified from the Anderson System (http://landcover.usgs.gov/classes.php), which is more appropriate to use from satellite imagery such as Landsat. Understanding the process by which LULC maps are created will provide insight into characteristics of photos or imagery that are useful in classification.

LEARNING OBJECTIVES

1. Explain convergence of evidence and how it is used in land-use/land-cover classification
2. Explain spectral reflectance and how it is used in land-use/land-cover classification
3. Understand the image classification process using supervised classification
4. Describe factors impacting accuracy in image classification

Elements of Image Interpretation and Classification

When interpreting features on imagery or using imagery to classify landscapes, it is necessary for one to be a detective of sorts. You need to observe several things within a landscape in order to correctly depict the earth's features. **Convergence of evidence** is a process whereby several elements are considered collectively to arrive at proper classification or identification of objects. There are several elements of image interpretation to consider: size, tone or color, shape, texture, pattern, shadow, site, and association. The **size** of an object, either absolute or relative, can reveal its identity. For instance, specific classes of objects such as trees and shrubs can be separated by size comparison. A regenerating forest contains smaller trees than a mature forest. The width of a tree crown may reveal species or forest types. Oak trees, for example, tend to have wide crowns, whereas aspen trees have narrow crowns. Coniferous trees such as spruce, fir, cedar, and some pine have small, pointy crowns, whereas deciduous trees have wider crowns. **Shape** refers to the form of an object. Shape may be more difficult to interpret from a vertical view. Shadows, however, may reveal the true shape of an object and aid in identification or classification of it. **Tone** or **color** refers to the variations in the reflectivity of different objects. For instance, conifer trees tend to be darker green than deciduous trees are. In winter, conifer forests are evergreen, whereas deciduous forests appear brown and striated. **Texture** refers to the coarseness or smoothness of an object or area. For example, corn fields appear rough-textured, especially in the fall when the plants have tassels. On the other hand, a wheat field appears smoother. Texture may be affected by scale, as detail is lost with smaller-scale photos. **Pattern** refers to the arrangement or placement of objects in an area. For instance, trees in an orchard have a regular distribution or pattern, whereas trees in a natural forest are randomly distributed. Areas such as golf courses have distinct patterns. **Site** may also aid in classification of features or objects within a landscape. Proximity to water or wetland areas, for instance, may be important in classifying forests as lowland versus upland. **Association** refers to the tendency of some objects to be associated with specific areas. Evidence of school buses in a parking lot may aid in classification of schools.

Reflectance

With land-cover classification, tone/color is probably the most important element used to classify areas. Tone or color may be used to identify different land-cover categories based on properties of reflectivity. Recall that reflection is the amount of electromagnetic energy that bounces off an object. In the case of imagery, we are referring to the reflectance of visible light or near infrared. Visible light can be separated into three bands: blue, green, and red. In the electromagnetic spectrum, infrared wavelengths are closest to the wavelengths of red visible light. In three-band imagery, only visible light is used in the acquisition of the imagery. In four-band imagery, the three visible light bands plus an infrared band is used in the acquisition of the imagery. The fourth band (infrared) provides more reflectance information about the land cover. In each case, however, different forest types or land-cover types reflect different amounts of blue, green, red, and/or infrared wavelengths. Computer software programs can measure the amount of reflectivity of the different wavelengths within each pixel on the image and categorize common areas. To illustrate this concept, notice in the pictures below that leaves from different tree species are not the same green. Oak and other hardwood leaves appear duller than the yellow-toned aspen or the pine. Thus, each species reflects a different proportion of the blue, green, red, and/or infrared wavelengths. This phenomenon is referred to as spectral reflectance. **Spectral reflectance** is the amount of the sun's radiation reflected by different land-cover types.

(a) Source: http://pixabay.com/en/oak-branch-leaves-acorns-fruit-418752/.

(b) Source: http://pixabay.com/en/leaves-road-branch-green-tree-357688/.

(c) Copyright © Keith Kanoti (CC by 3.0 US) at https://commons.wikimedia.org/wiki/File:Pinus_resinosa_cone.jpg.

Techniques for LULC Classifications

Two primary techniques are frequently used in LULC classifications: supervised and unsupervised classification. In the process of supervised classification, an analyst (i.e., you) delineates **training samples** in imagery around groups of pixels within known land-cover types. Computer software programs designed for land-cover classification assess the range of reflectance values characterized by the training samples, and then classify the rest of the image based on similarities in reflectance to the training samples. In unsupervised classification, computer software first separates the image pixels into natural groupings based on similar spectral characteristics. Then, the classification analyst (you) assigns each grouping to a land-cover class. ERDAS Imagine® and ArcGIS® are two examples of software frequently used in the process of land-cover classification.

The process of supervised classification is illustrated in the four-band imagery below. Across a landscape, the difference in spectral reflectance among different land-cover types is evident. These differences allow us to create LULC maps from imagery. For instance, it is clear that red pine appears a dark magenta or purple color, whereas aspen appears orange-ish. Mixed hardwood reflects more pink tones, but there is much variation depending on species. The dotted lines on the image indicate training samples. If you were conducting a supervised classification, you might first use convergence of evidence to identify areas of different vegetation types. For instance, red pine is frequently planted in rows within plantations on dry upland sites. Therefore, the regular **pattern** of pixels that appear as a magenta or purple **tone** in areas you know are **associated** with dry upland sites would help you identify areas that should be classified as red pine. So, after identifying such areas within the image, you can delineate a training sample (depicted with the dotted line in the following image) to "train" the software to recognize red pine throughout the rest of the image. On the other hand, aspen trees grow very closely together in clones, and their canopies are relatively narrow compared to other hardwood species. Knowledge of the **pattern** of aspen growth, the size of the canopy, and the **tone/color** of reflectance allows correct image interpretation and subsequent classification. Once again, training samples are used to train the software to recognize the range of reflectance values associated with aspen. In the following histogram, it is easy to see the difference in reflectance between the darker red pine and the brighter aspen. Pine reflects different wavelengths on the electromagnetic spectrum than aspen does. The index of reflectance on the x-axis is color values depicting different wavelengths reflected (created in ArcGIS® v 10.1). Land-cover classification software programs use those quantitative differences in reflectance in the process of classification.

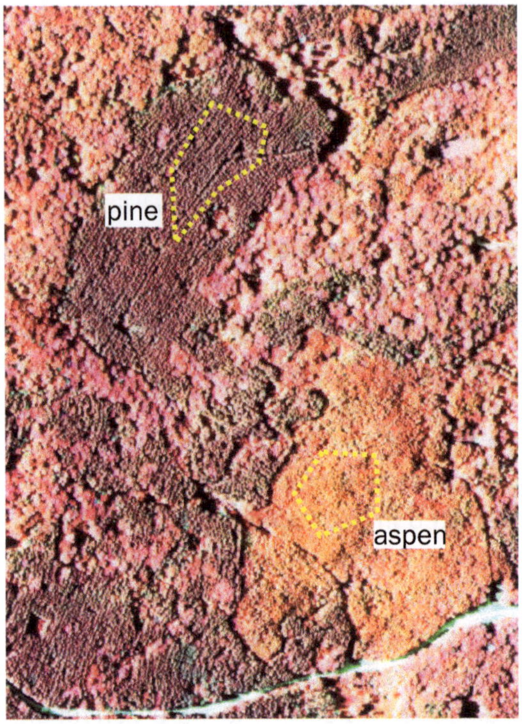

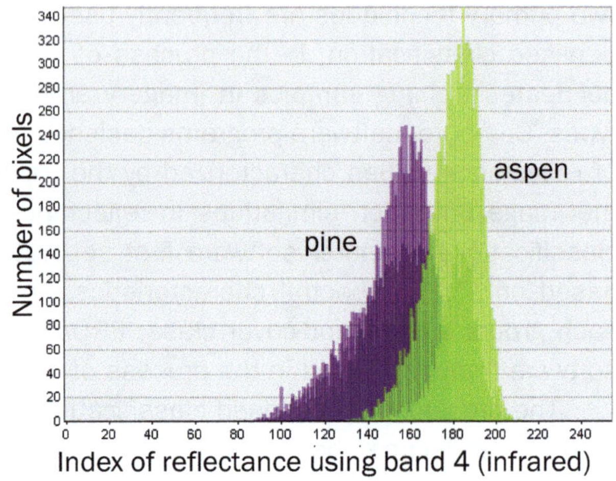

(a) Source: National Agriculture Inventory Program, U.S. Department of Agriculture.

(b) Copyright © by Esri.

Once you have delineated several training samples for each land-cover type, you can run the tools within the classification software program to create the final land-cover classification maps. In the following images, compare the raw data on the left with the classified image on the right. Can you see how pattern, tone/color, and other elements of image interpretation can be used to create a map depicting LULC?

1 x 1 meter aerial imagery

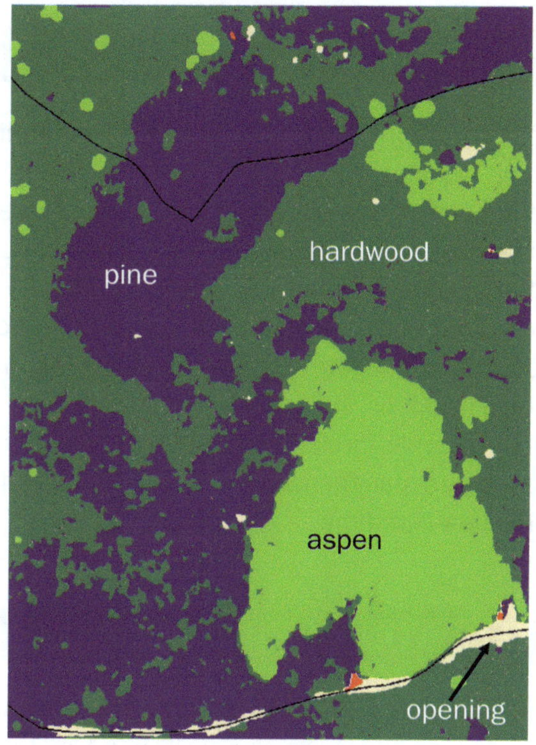

Classified imagery

(a) Source: National Agriculture Inventory Program, U.S. Department of Agriculture.

Factors Affecting Classification Accuracy

Several factors affect the degree of accuracy in land-cover classification. First, classification requires extensive knowledge of reflectance, patterns, and other characteristics associated with different land-cover types. Uncertainty from novice analysts may affect accuracy. Second, flaws or obstructions in the initial images such as cloud cover, haze, or seams in mosaicked imagery may affect accuracy. Third, shadows may be incorrectly classified as water or other land-cover types that commonly reflect darker tones. Fourth, wide variability in reflectance in land cover such as agriculture may result in misclassification. Software tools, experience, and skill can all help minimize error in classification.

SUMMARY

Interpretation skills are important in identifying objects in aerial photos and determining their significance; they are also helpful in classifying images in terms of land use and land cover. When classifying imagery, several elements are considered collectively to arrive at a proper classification (i.e., convergence of evidence). These elements may include shape, tone or color, texture, pattern, site, or association. To classify land cover types, computer software programs may be used to distinguish among spectral reflectance values for different vegetation types or surfaces on the earth. These software programs may use supervised or unsupervised classification techniques. Land use/land cover classifications provide fundamental information about the earth and are frequently used in natural resources-management fields or earth science fields to understand current landscapes and changes over time.

TEST YOUR SKILLS

Use elements of image interpretation (i.e., convergence of evidence) to classify areas specified in the images below. Indicate what element(s) of interpretation helped you correctly classify the areas according to the following classification scheme:

Commercial:	highly developed areas characterized by ≥50% construction materials
Residential:	areas with a mixture of residential structures and vegetation
Park/golf course:	mostly vegetated areas maintained for recreation
Agriculture:	areas maintained for row crops, grains, or pasture
Orchard/vineyard:	areas planted for the production of food or nursery stock
Deciduous forest:	≥60% of an area composed of deciduous trees (i.e., trees that shed leaves in the fall)
Coniferous forest:	≥60% of an area composed of evergreen trees
Mixed forest:	forested area with approximately 50% deciduous and 50% coniferous trees
Wetland:	vegetated areas sustaining water for all or part of the year
Water:	all areas of open water

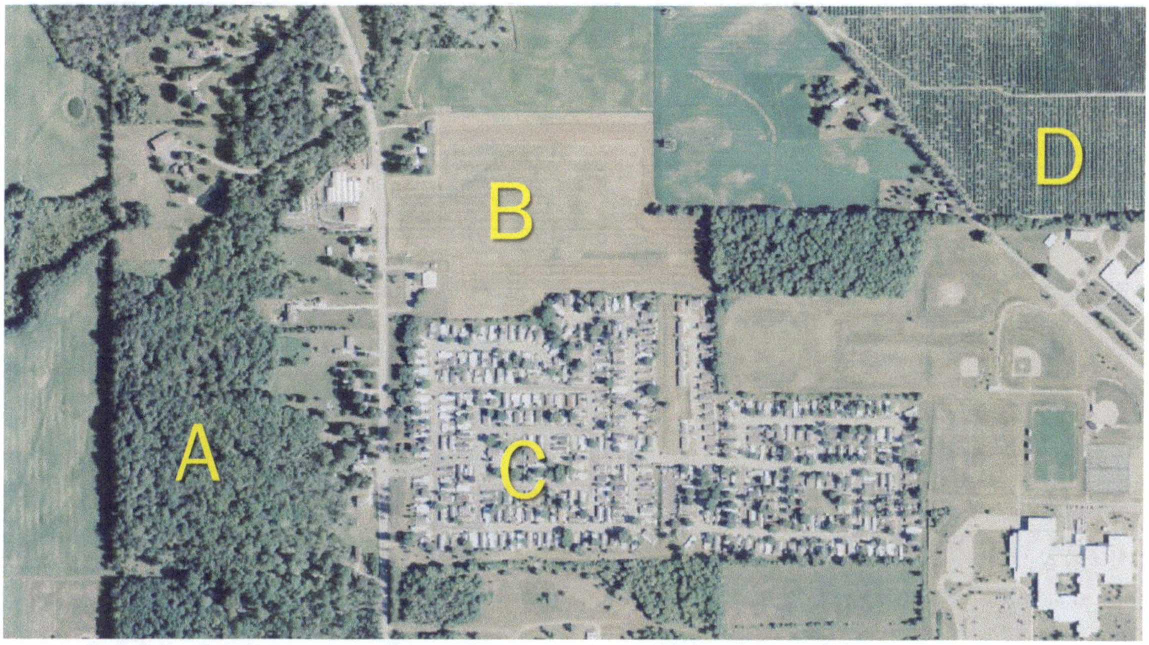

Source: National Agriculture Inventory Program, U.S. Department of Agriculture.

Area	Classification	Element(s) of Interpretation
A		
B		
C		
D		

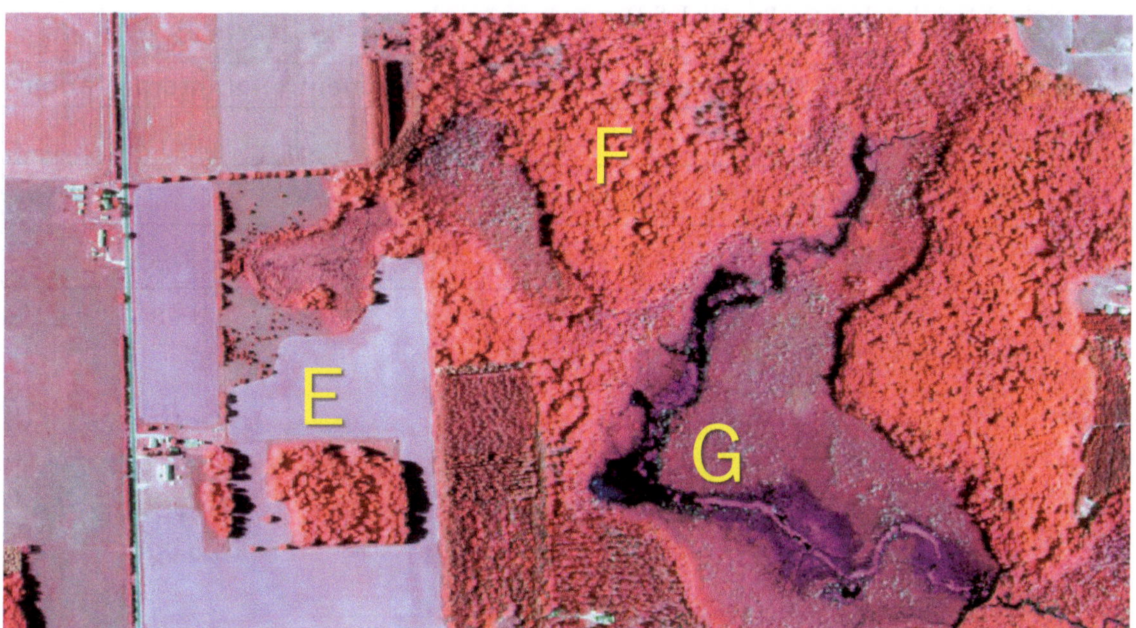

Source: National Agriculture Inventory Program, U.S. Department of Agriculture.

Area	Classification	Element(s) of Interpretation
E		
F		
G		

Source: National Agriculture Inventory Program, U.S. Department of Agriculture.

Area	Classification	Element(s) of Interpretation
H		
I		
J		

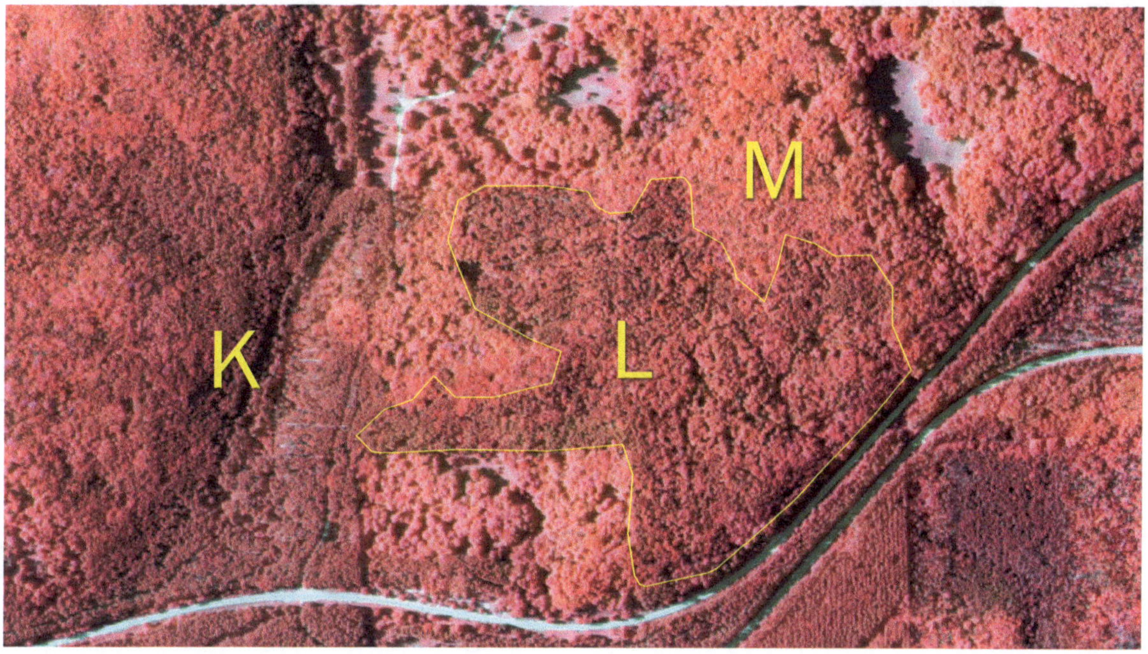

Source: National Agriculture Inventory Program, U.S. Department of Agriculture.

Area	Classification	Element(s) of Interpretation
K		
L		
M		

REFERENCES

Aronoff, S. (2005). *Remote Sensing for GIS Managers*. Redlands, CA: ESRI Press.

Enderle, D. M. & Weih, R. C., Jr. (2005). Integrating Supervised and Unsupervised Classification Methods to Develop a More Accurate Land Cover Classification. *Journal of the Arkansas Academy of Science*, *59*: 65–73.

Paine, D. P. & Kiser, J. D. (2003). *Aerial Photography and Image Interpretation*, 2nd Edition. Hoboken, NJ: John Wiley & Sons.

12

Global Positioning Systems

Global Positioning Systems (GPS) are used widely in natural resources professions and in everyday life. Thanks to the US Department of Defense, GPS satellites provide us with the capability to locate position and find destinations with readily available and affordable receivers. GPS is a satellite navigation system initially designed for and operated by the military, but is now available to everyone worldwide. GPS provides specially coded satellite signals that can be processed in a hand-held receiver. The receiver is able to compute position, velocity, and time.

LEARNING OBJECTIVES

1. Explain the evolution of navigation technology
2. Understand the principles of GPS
3. Recognize the use of latitude–longitude coordinates in navigation
4. Understand sources of positioning error when using GPS

Evolution of Navigation Technology

To understand the concept of navigation with GPS, it may be helpful to review the evolution of steps that led up to GPS technology. Long ago, early humans assessed their position and navigated using natural landmarks or stone piles as points of reference. A **point of reference** in navigation is a stationary object that helps you determine your position and orient yourself in a particular direction. For instance, a benchmark used in surveying is a point of reference. Benchmarks are located in permanent positions and are used in the determination of other vertical and horizontal positions in boundary or topographic surveys. Other common objects are frequently used as points of reference every day. A buoy, for example, marks channels in a harbor or underwater hazards to facilitate marine navigation. An intersection at two familiar crossroads can indicate how much farther it is to your destination.

As early as several centuries BC, humans used celestial bodies such as the sun, moon, and stars as points of reference for navigation. Celestial navigation was used for thousands of years. Specific stars such as the North Star, or constellations such as the Big Dipper, Draco the Dragon, and Orion were commonly used to determine one's position and facilitate navigation.

Invention of tools such as the **astrolabe** and the **sextant** paved the way for early earth explorers to navigate across oceans. Such tools measured angles of celestial bodies relative to the horizon and allowed determination of latitude. Recall that latitude is determined by subtracting the angle of the sun by 90° and then adjusting the difference by the analemma correction based on the sun's declination on that particular day (see Chapter 1). Invention of the **chronometer**, an extremely accurate clock, allowed calculation of time relative to a point of reference such as the Prime Meridian. This invention allowed calculation of longitude.

Celestial navigation was used for centuries until technological developments in the twentieth century used ground-based radio waves for navigation. Use of ground-based radio waves to determine position is a form of active remote sensing (as opposed to passive remote sensing; see Chapter 10). In this process, a radio signal is generated from a transmitter. A receiver detects the signal and calculates the distance from the transmitter by simple physics: distance = rate of transmission * time it takes to receive the signal from transmitter. It takes three transmitters to triangulate position. The transmitters involved in the calculation of location are called a transmitter "chain."

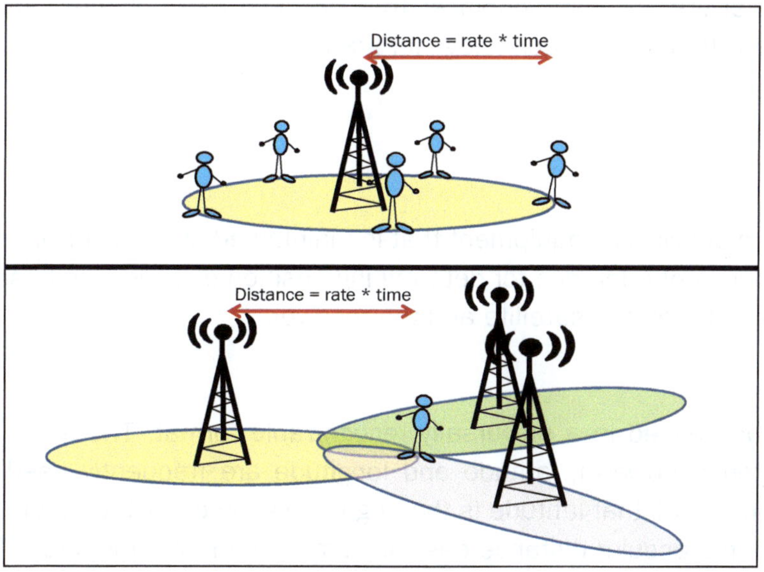

LORAN (Long Range Navigation) was active from 1950 to 2010. It was developed in the United States during World War II, and improved positioning accuracy in ocean navigation. LORAN was not a global system; its use was limited primarily to the military. In 2010, the LORAN system was decommissioned because GPS was more widely used.

The first satellite-based navigation system was called TRANSIT. It was launched and tested by US military in 1960. GPS technology was used in the military after the launch of the NAVSTAR satellite in the 1970s, but it wasn't until the 1990s that it became fully operational for the military and available for civilians.

Segments of GPS

There are three segments to GPS: space, control, and user. The space segment consists of more than twenty-four satellites orbiting 12,000 miles above the earth. Satellites are configured in six orbital planes. Each satellite orbits the earth twice within a 24-hour period, and repeats the same track approximately every 24 hours. They are arranged so that five to eight are visible at any point. Satellites travel at 7,000 miles per hour, which would be equivalent to traveling from the eastern United States to central China within 1 hour. Satellites weigh more than 2,000 pounds, and are approximately 17 feet across with solar panels extended. They cost tens to hundreds of millions of dollars to build, launch, and operate. The control segment involves a system of tracking stations located around the world. The Master Control station is located at Falcon Air Force Base in Colorado Springs, Colorado. The control stations collectively have contact to every satellite at any given time and can monitor them. The user segment contains the GPS receivers and the user community. GPS receivers are commonplace; they occur in cars, phones, and other mobile devices. It is easy to take this technology for granted.

Principles of Satellite Navigation

Satellites

Satellites require specialized equipment that is maintained by the control segment of GPS. They serve as points of reference for determining position of a receiver based on computation of distance between the satellite and the receiver.

Coordinate system

Position must be defined in a universally recognizable format. There are many types of coordinate systems; however, latitude and longitude are frequently used within GPS to indicate position. Recall that latitude is the angular distance north or south of the equator, and longitude is the angular distance east or west of the Prime Meridian. There are three primary ways to indicate latitude and longitude: degrees-minutes-seconds, degree minutes, or decimal degrees.

Degrees-minutes-seconds – Degrees may be divided into 60 minutes, and each minute may be divided into 60 seconds. Thus, positions between lines of latitude or longitude may be expressed as the number of degrees, the number of minutes, and the number of seconds. The distance beween lines of longitude is approximately 69 miles at the equator. Thus, the distance between one minute of longitude is 1.15 miles, and the distance between 1 second of longitude is about 101 feet. The distance between longitude varies with position because lines of longitude converge at the poles. For example, distance between lines of longitude on the equator is approximately 69.17 miles. Distance between lines of longitude at 45° latitude (halfway between the equator and the north pole) is roughly 49 miles. The Spherical Law of Cosines is a complex equation used to calculate distance between lines of longitude at varying latitudes, but you can reasonably estimate distance with the following equation:

$$\textit{Distance betwen lines of longitude (miles)} = \textit{cosine}(\textit{latitude}) * 69.17 \textit{ miles}$$

In any case, coordinates reported in degrees-minutes-seconds appear like the following:

N 42° 45' 53.64" North 42 degrees, 45 minutes, and 53.64 seconds

W 85° 30' 10.08" West 85 degrees, 30 minutes, and 10.08 seconds

Degree-minutes – Coordinates formatted in degree-minutes convert seconds to a fraction of a minute by dividing by 60. Then, add the fraction to the minutes. For example, let's convert the coordinates above to degree-minutes.

Deg-min-sec	Conversion	Degree-minutes
N 42° 45' 53.64"	42° 45 + (53.64 / 60)'	N 42° 45.894'
W 85° 30' 10.08"	85° 30 + (10.08 / 60)'	W 85° 30.168'

Decimal degrees – To format coordinates using decimal degrees, divide the decimal minutes by 60 in the example above, and add it to the degrees.

Deg-min-sec	Conversion	Degree-minutes
N 42° 45.894'	42 + (45.894 / 60)°	N 42.7649°
W 85° 30.168	85 + (30.168 / 60)°	85.5028°

Determination of position

The distance between satellites and a receiver is necessary to triangulate position. First, the receiver connects to satellites once it is powered on. Many handheld receivers indicate which satellites they are receiving signals from. Connection to satellites is based on line of sight. In open areas, a receiver may connect to more satellites than it can when it's in a valley bounded by high hills. In a forest, the tree canopy may obstruct connection between a receiver and satellites. Once the receiver has connected to satellites, the distance between the receiver and the satellite is calculated. We know that distance is equal to rate times time. In GPS, rate refers to the speed at which a special code (pseudo-random code) is emitted from a satellite. This code travels at the speed of light, which is equivalent to something traveling around the world 7.5 times in 1 second! Time refers to the time delay between when the code was emitted from the satellite and when it was received. The position of the satellite is determined through ephemeris, a decoded message in the satellite. Horizontal position is triangulated from three satellites; a fourth satellite is needed to compute time and elevation. Recall that triangulation is a method of determining a position by measuring the distances between that position and other points of reference (satellites, for example).

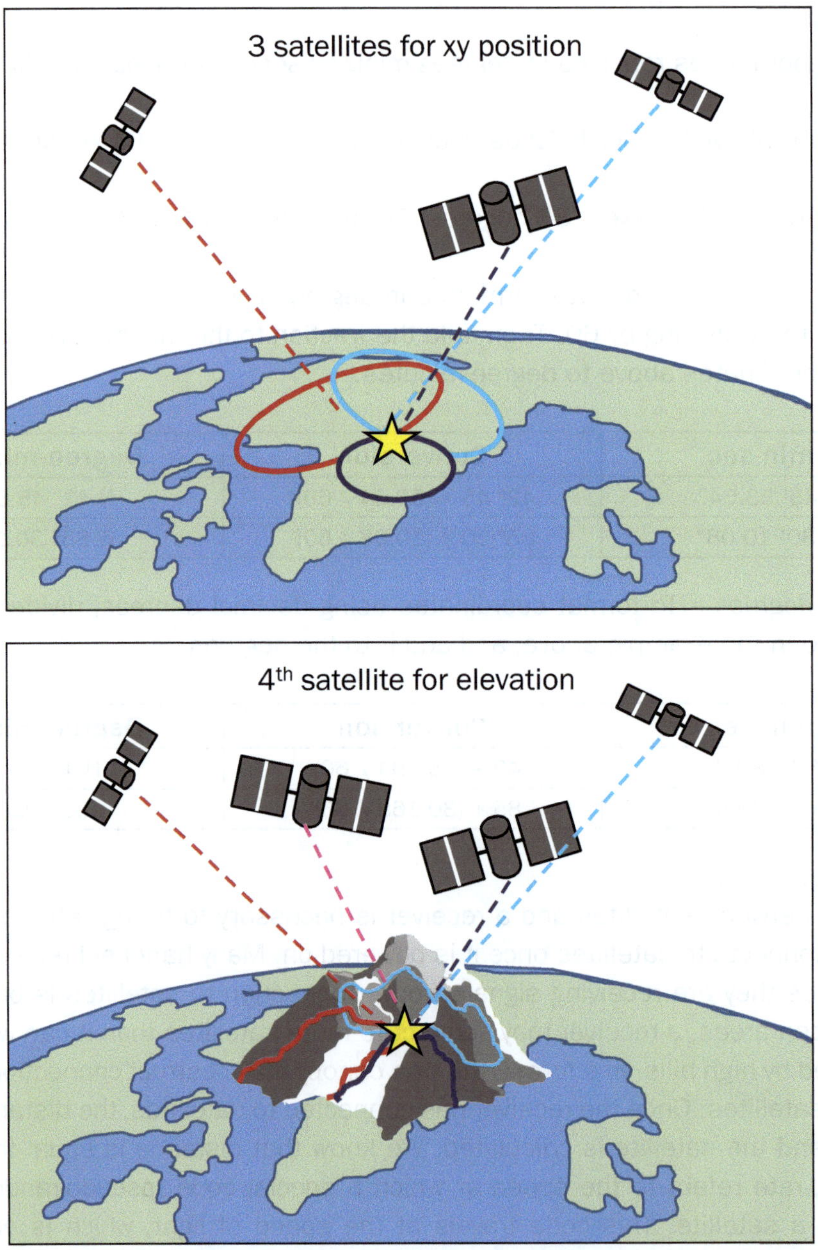

GPS sources of error

The accuracy of positioning with GPS varies with type of receiver (i.e., recreational grade, mapping grade, or survey grade), and make/model. Although accuracy is generally very good, there are some sources of positioning error that you should be aware of. First, atomic clocks within the GPS satellites accumulate an error of 1 nanosecond every three hours, which can cause errors up to 30 cm in measuring distance. Second, the receiver clock contains a common quartz clock that can drift 1,000 nanoseconds every second. This quartz clock is constantly resynchronized with the satellite clock to adjust for the time error. Third, there may be some error in determining the precise location of the satellites. Because the satellites act as points of reference, if their position is biased, distance calculations and position of the receiver may also be biased. Fourth, the atmosphere and the environment may cause positioning errors in GPS. Charged particles, water vapor, dust, or other obstructions such as tree canopy cover or buildings may interfere with the connection between satellites and receivers. Fifth, a phenomenon referred to as "multipath" occurs when the signal from the satellite bounces off other objects and is subsequently detected by the receiver. The receiver "thinks" the signal is coming from a different direction and thus, results in positioning bias. Finally, **geometric dilution of precision** may affect positioning accuracy. Dilution of precision occurs when the satellites are very close together (i.e., all satellites are overhead). Positioning accuracy is not as good as if the satellites were further apart (i.e., dispersed between the horizon and overhead).

Differential correction

Positioning errors may need to be corrected. Differential correction involves obtaining a correction factor for GPS errors by comparing the known position of a base station with its calculated position from GPS satellites. The difference between the actual position of the base station and the calculated position of the base station is the GPS error. This error is computed repeatedly throughout the day. Users can use the error computations from the base station to adjust for their positioning errors with mobile receivers. There are computer software programs to differentially correct position or programs that automatically differentially correct position in mobile receivers, but the process is illustrated on the next page.

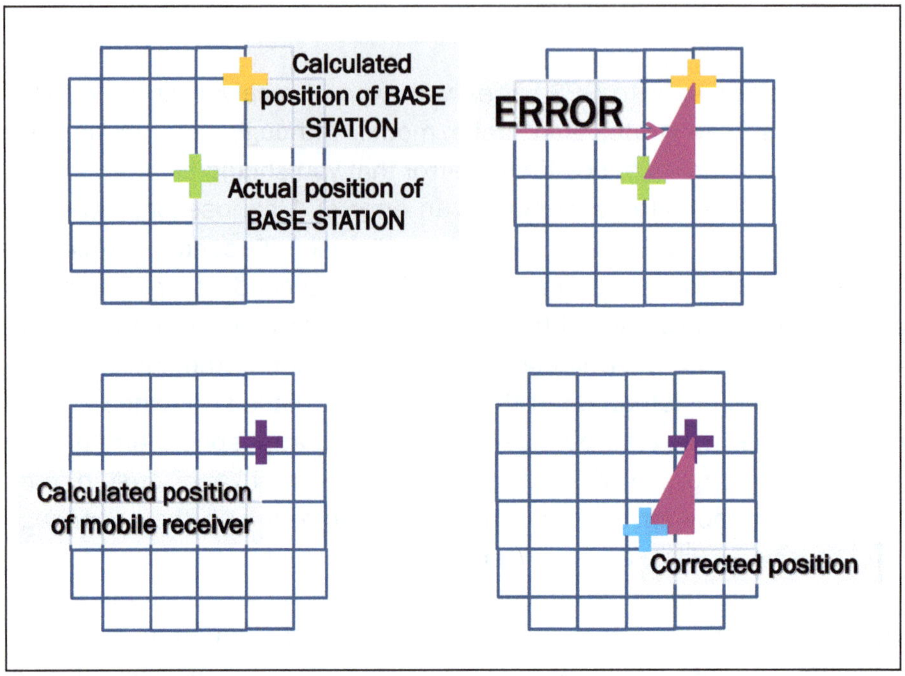

► For example, your GPS receiver calculated a position with the following coordinates: N 44.34628°, W 92.76831°. Suppose differential correction indicated the positioning error was 150 feet SOUTH and 150 feet WEST of the actual position. What are the corrected coordinates? [Note: the positioning error in these problems is much larger than the error you would likely get on your receiver. The problems are intended to illustrate the correction process.]

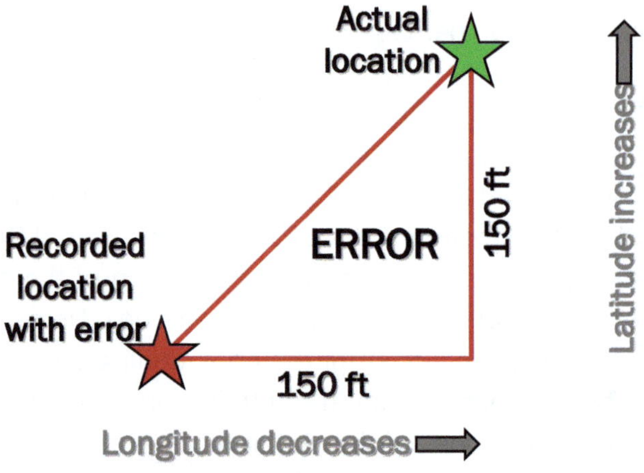

1. To correct the coordinates, we must find the degrees equivalent to the error of 150 ft. First, it may be helpful to convert the error to miles, because it is easier to convert between degrees and miles versus degrees and feet.

$$\frac{150\ \cancel{ft}}{1} = \frac{1\ mile}{5{,}280\ \cancel{ft}} = 0.0284\ mi$$

2. Now we can set up a ratio between degrees of latitude and distance in miles.

$$\frac{1\ degree\ latitude}{69\ miles} = \frac{x}{0.0284\ miles}$$

$$x = 0.00041\ degrees$$

So, we know that the error in latitude is 0.00041 degrees. Because the actual location is NORTH of the recorded erroneous location, we must ADD the difference to the original recorded latitude.

$$N\ 44.34628° + 0.00041° = \mathbf{44.34669°}$$

3. To correct for longitude, we still must find the degrees equivalent to the error of 150 ft, but the calculation is not as straightforward because the distance between lines of longitude changes between the equator and the poles. To estimate the distance between lines of longitude at a specific latitude, take the cosine of latitude and multiply by 69.17, which is the distance in miles between lines of longitude at the equator.

$$\cos(44.34669) * 69.17 = 49.4651\ miles$$

Next, set up a ratio between degrees of longitude and distance in miles.

$$\frac{1\ degree\ longitude}{49.4651\ miles} = \frac{x}{0.0284\ miles}$$

$$x = 0.00057\ degrees$$

Finally, correct the longitude coordinate. We know that the actual position is EAST of the erroneous position. Therefore, we must SUBTRACT the error in longitude (because longitude increases from east to west).

$$W\ 92.76831° - 0.00057° = \mathbf{92.76773°}$$

The corrected position is N 44.34669°, W 92.76773°.

SUMMARY

Navigational techniques have evolved over time. Throughout history, determining position has changed from using natural landmarks and celestial bodies as points of reference, to complex technological advancements such as satellites and computers. Global Positioning Systems (GPS) use more than 24 satellites, a centralized control system operated by the US Department of Defense, and a user-centered receiver to compute position. GPS capabilities are widely available to people via commercial receivers or mobile phone devices. GPS defines position in a universally recognizable format (i.e., latitude and longitude), but may also project positions in other coordinate systems. Although GPS positioning is fairly accurate, sources of error that degrade the precision of location estimates include: drift in the satellite atomic and receiver clocks, drift in satellite position, obstructions in the atmosphere or environment, and signal bounce. Differential correction, however, can be used to correct positioning errors. GPS are used widely in natural resources professions and in everyday life.

TEST YOUR SKILLS

1. Convert the following positions in decimal degrees to degrees-minutes-seconds format. Keep two decimal places for the seconds.

   a. N 45.60978° __________ W 92.00992° __________

   b. N 32.76001° __________ W 75.18622° __________

   c. S 21.99802° __________ E 12.45463° __________

2. Convert the following positions in degrees-minutes-seconds to decimal degrees. Keep five decimal places.

   a. N 50° 34' 32.15" __________ E 25° 14' 58.49" __________

   b. S 10° 14' 2.11" __________ E 100° 54' 55.30" __________

   c. S 50° 29' 8.41" __________ W 31° 25' 57.99" __________

3. Using the diagram to the right, what is the distance between A and B in miles?

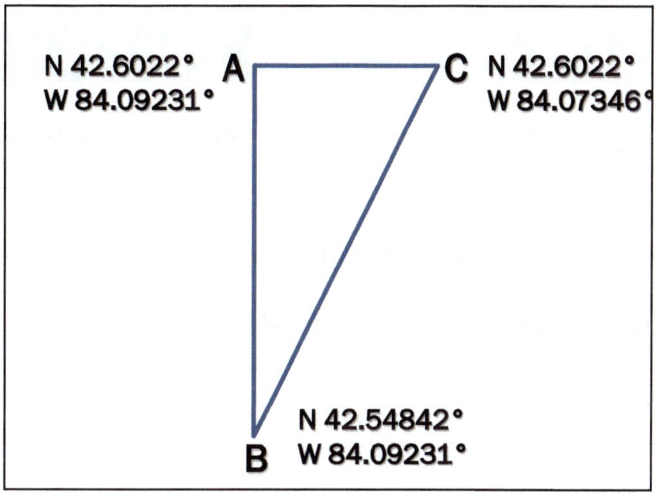

4. Using the same diagram, what is the distance between A and C?

5. Using the same diagram, determine the distance between B and C.

6. Suppose differential correction indicated that your positioning error was 115 feet too far north and 97 feet too far east. If your starting coordinates were N 42.54842 W 84.09231, what are your new coordinates?

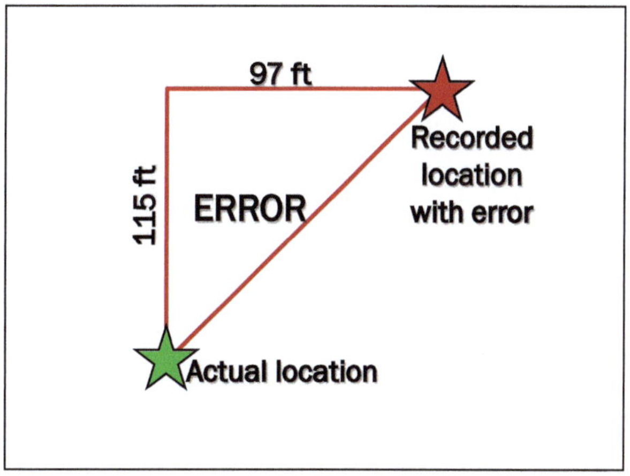

7. Practice navigating with a GPS. Geocaching is a fun way to learn navigation with GPS. Have someone hide a treasure and give you the coordinates, or find geocaching locations online. Navigate away!

REFERENCES

Avery, T. E. & Burkhart, H. E. (2002). *Forest Measurements*, 5th Edition. New York, NY: McGraw-Hill.

Hurn, J. (1993). *Differential GPS Explained*. Sunnyvale, CA: Trimble Navigation Limited.

Kennedy, M. (2009). *The Global Positioning System and ArcGIS*, 3rd Edition. Boca Raton, FL: CRC Press.

13

Using GIS for Spatial Data Analysis and Display

The initialism "GIS" stands for Geographic Information Systems. As you begin to search for jobs in natural resources–related fields, you may be learning that GIS skills are listed as a requirement or a preferred qualification in many of the advertised positions. Geographic Information Systems integrate computer hardware and software for capturing, storing, analyzing, and displaying spatial data. An understanding of spatial analyses including GIS will increase your marketability for future jobs. The uses for and applications of GIS are endless: you can use GIS to identify forest stands for timber harvest, compare landscape changes over time, quantify sedimentation risk to trout streams along banks with high slope, interpolate temperatures, create maps depicting habitat suitability, and many others.

GIS is a *system*; thus, it consists of several things. First, GIS requires computer hardware and software. It requires spatial data to be processed. Spatial data may be aerial imagery, GPS points, files depicting land cover or other features on the earth, files depicting population density or political boundaries, and many other types of data. Finally, GIS requires some sort of infrastructure including skills, data standards, organization, and users like you!

Spatial data have two characteristics: attributes and location. **Attributes** are characteristics about the specific entity with which you are working. For instance, if you were studying lakes, some attributes of lakes could be size, perimeter length, water temperature, depth, number of houses along shore, or water quality. Attributes allow you to ask *what* questions about entities on the earth. **Location** refers to the coordinates or position of each feature. Location allows you to make inferences about a feature's position relative to other entities on the earth. Location allows you to ask questions about *where* specific conditions are met. GIS is much more than a graphic arts program or a mapping program. In essence, the power of GIS lies with the capability to conduct spatial analyses.

The purpose of this chapter is to introduce you to GIS and help you begin to understand the capabilities of GIS. You will learn a very basic and broad overview of some fundamental GIS concepts.

LEARNING OBJECTIVES

1. Describe the two characteristics of spatial data
2. Explain GIScience and spatial thinking
3. Define *spatial data model* and distinguish between vector and raster
4. Describe entity–object correspondence
5. Explain how a thorough understanding of coordinate systems is important in GIS
6. Discuss applications of GIS to natural resource–related fields

Spatial Thinking

GIS has been used since the late 1960s to process geospatial data. However, computers used to process such data at that time were very large and complex. It has been only since the 1990s that GIS software developments and computer technology advanced such that GIS could be used in a relatively affordable graphical user interface for mainstream applications. Subsequently, **GIScience** (Geographic Information Science) emerged as an accepted discipline. GIScience seeks to better understand the nature of geographic data and how it may be collected and analyzed, using theories and logic from other disciplines such as cartography, surveying, remote sensing, geography, mathematics, and statistics. Essentially, GIScience is a discipline that uses GIS as a tool to better understand the earth and processes occurring on it. GIScience requires the ability to think spatially. The term **spatial thinking** has emerged and evolved along with the development of GIScience. Spatial thinking is a thought process that is used to solve many geographic problems. It interprets meaning in the attributes, location, shape, and orientation of entities on the earth. And if you think about it, everything is geographic. That is, everything can be defined, in part, by location. So if we are able to think spatially, and if we are able to use tools such as GIS to better understand aspects of our world, we can make better decisions about how we interact with our environment to maintain, conserve, and sustain resources on which we depend.

For example, suppose a fish biologist noticed a decline in trout in one part of a stream network. The habitat structure seemed suitable in the stream segment where the decline was occurring, except a layer of silt was accumulating over clean gravel necessary for viable trout eggs. A spatial analysis of the entire stream network within the watershed revealed that careless land management along steep slopes adjacent to the stream had decreased vegetative cover and subsequently increased erosion, which resulted in accumulation of silt downstream. GIS is a tool that can help the biologist identify areas with high-priority needs for streambank stabilization. Without spatial thinking and analysis using spatial data and tools, it would be difficult to locate the source of the sedimentation. The source(s) of sedimentation can be described spatially in terms of distance or direction from the declining trout population.

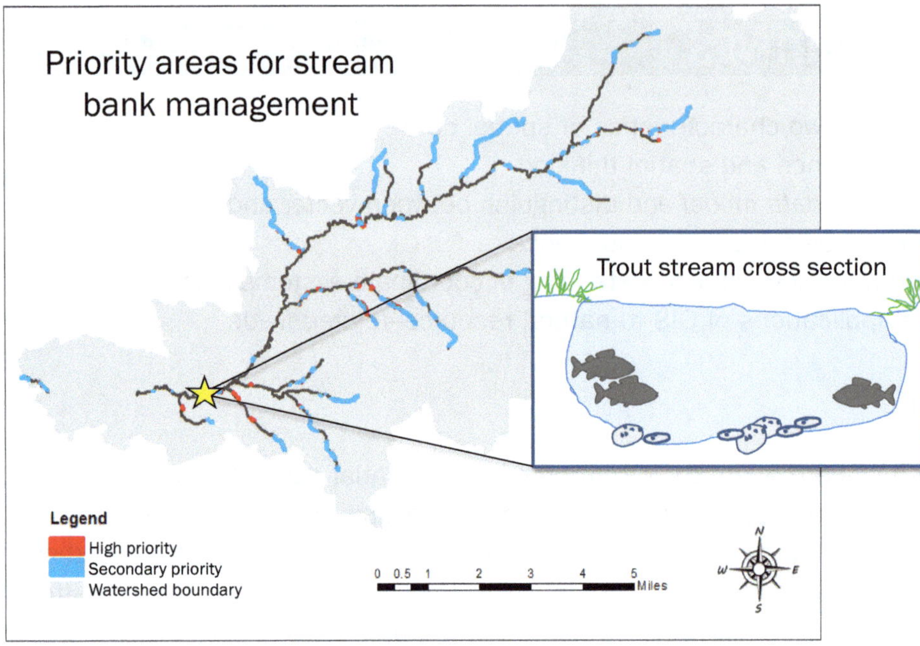

Spatial Data Models

Entity-object correspondence is a term frequently used in GIS to describe the relationship between features on the earth and how they are represented on maps. Essentially, an entity is something that exists. It could refer to any feature or phenomenon in existence on the earth. For example, lakes are entities; buildings are entities; countries are entities. In spatial data, entities are represented by cartographic objects such as points, lines, polygons, or pixels. A lake, for instance, can be represented by a polygon. Roads may be represented by lines. The polygons or lines in these examples are the objects that are used to represent the lake and road entities.

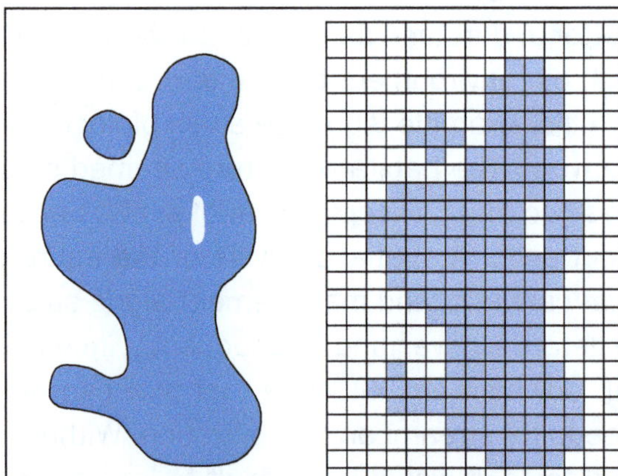

Representation of a lake using a vector data model

Representation of the same lake using a raster data model

Spatial data models define the type of object used to represent entities. Their purpose is to provide a formal means of representing and processing spatial data. Spatial data models can be classified as vector or raster. **Vector** data models represent features as points, lines, or polygons. **Raster** data models represent aspects of the world as pixels or grid cells.

Location

Recall that the two characteristics of spatial data are location and attributes. Location is defined by **coordinates**. Coordinates define location in two- or three-dimensional space. You are familiar with one universal coordinate system: latitude and longitude. Latitude defines position north or south of the equator. Longitude defines position east or west of the Prime Meridian. This coordinate system is referred to as a **geographic coordinate system**, or one that defines angular measurements on a sphere (e.g., the earth). Another type of coordinate system is the **Cartesian coordinate system**, which defines position using right-angle x- and y-axes, or x-, y-, and z-axes if elevation is included. Cartesian coordinate systems are created through projections to minimize distortion in shape, area, distance, or direction that may occur by flattening large geographic areas on the earth, which are curved due to being on the surface of a spheroid.

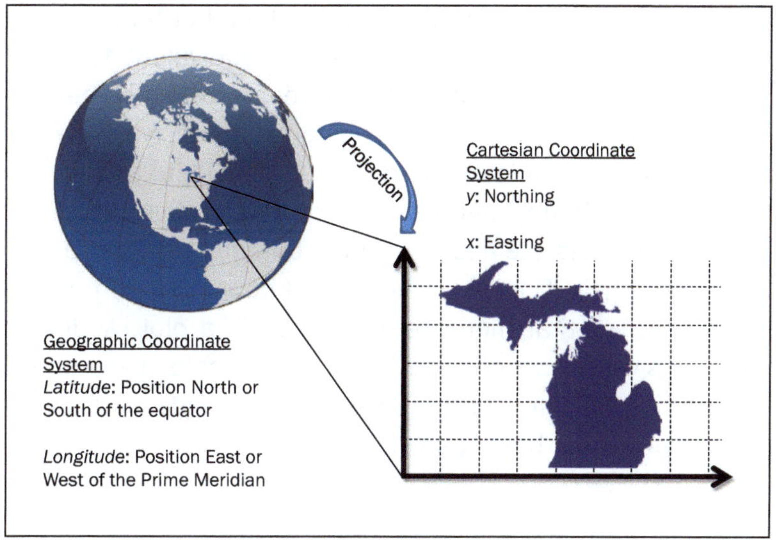

(a) Source: https://pixabay.com/en/globe-earth-planet-continents-147715/.
(b) Source: https://pixabay.com/en/michigan-map-state-america-23565/.

Attributes

Attributes are characteristics of an entity. For example, if you were analyzing roads, attributes might be "road name," "road length," or "traffic volume." If you were analyzing the distribution of hunter-harvested deer, attributes might be "individual ID," "sex," "number of antler points," or "antler beam diameter." In essence, attributes provide more information about an entity than location alone.

Attributes can be classified into four different types: nominal, ordinal, interval, and ratio. It is important to distinguish among different types of attributes because queries and appropriate spatial analyses differ depending on attribute type. **Nominal attributes** provide descriptive information about an entity. Attributes such as "name," "color," "ID," or "species" are all nominal attributes (Table 13-1). **Ordinal attributes** imply ranking or order. They may be numeric, or they may be represented as text. An example of an ordinal attribute is habitat suitability represented as a number between 0 and 1; 1 indicating high suitability and 0 indicating low suitability. An index of water quality ranked as high, medium, and low is another example of an ordinal attribute. **Interval attributes** are those that have a defined distance between values. For instance, time of day, calendar date, temperature, or latitude–longitude all have specific distances between successive intervals, and a value of 0 does not indicate "nothing." With interval data, the zero point is arbitrary. For instance, 0° Fahrenheit and 0° Celsius are not equivalent temperatures. If you are working with calendar date, year 0 could be any year in which something started. A latitude of 0° does not indicate "no location." The final attribute type is ratio. **Ratio attributes** have values where the magnitude of difference is meaningful and can be added, subtracted, multiplied, and divided. Ratio attributes cannot be negative; thus, if the value of a ratio attribute is zero, it indicates the absence of the quantity being measured. For instance, diameter, area, perimeter, height, length, weight, and count are all ratio attributes. A count of 0 trees in a plot means that there were no trees in a plot. Additionally, you can compute ratios or fractions with this type of variable. For example, if you were measuring tree diameter, a tree measuring 30 cm in diameter is twice as wide as a tree measuring 15 cm in diameter; diameter cannot be negative.

Table 13-1 Descriptions and examples of four types of attributes

Attribute type	Description	Examples
Nominal	Provide descriptive information	Name, color, sex, type, species, ID
Ordinal	Ranks or orders	Small–large, 1–10 (10 = high), colder–warmer
Interval	Have a defined distance between values and zero value is arbitrary	Time of day, year, latitude–longitude, temperature
Ratio	Zero values indicate absence	Height, weight, count, area, perimeter, length, diameter

Types of Spatial Analyses

The possibilities of analyses using GIS are endless! Jack Dangermond, the founder of Environmental Systems Research Institute (the company that produces ArcGIS® software), said that "the application of GIS is limited only by the imagination of those who use it." Below are some basic categories of applications of GIS to help you start understanding the types of analyses for which GIS may be used.

Attribute and location queries

Suppose that you were a wildlife biologist studying the distribution of woodland jumping mice to understand their habitat use patterns. So you trapped mice in different forest types and recorded their presence using a GPS receiver. If you wanted to understand if there were specific vegetation types used by the mice more than other vegetation types, you could conduct queries based on the location of points in relation to vegetation types. For instance, you could query the vegetation types in which a mouse location occurred and then tally the frequency of mouse locations *within* each vegetation type. The relationship "within" is an example of a spatial relationship based on containment. You might also decide to query the number of locations occurring at a specific distance to another feature such as a stream, or you could also conduct location queries based on intersection or adjacency. Additionally, if your data contained information on date of capture, you could query the number of mouse locations during different months to understand if habitat use differed seasonally.

Point patterns

Working with point data is extremely common in natural resources–related fields. For example, some common point data are locations of nests, invasive plants, wildlife or fish tracked with radio transmitters, sample points, deer–vehicle collisions ... the list is endless. Sometimes, it is important and informative to analyze the patterns of point distributions. Point-pattern analysis quantitatively determines whether the distribution of points is random, clumped, or dispersed. Such information allows inferences to be made for management. For instance, suppose you were in charge of assessing whether biological control implemented for an invasive wetland plant, purple loosestrife, was effective. So, you designed a study to compare the spatial distribution of purple loosestrife before and after treatment. If you determined that the spatial distribution of the invasive plant was relatively dispersed within the study area before treatment and clumped in smaller groupings after treatment, you might conclude that the treatment was effective in reducing the distribution of the plant.

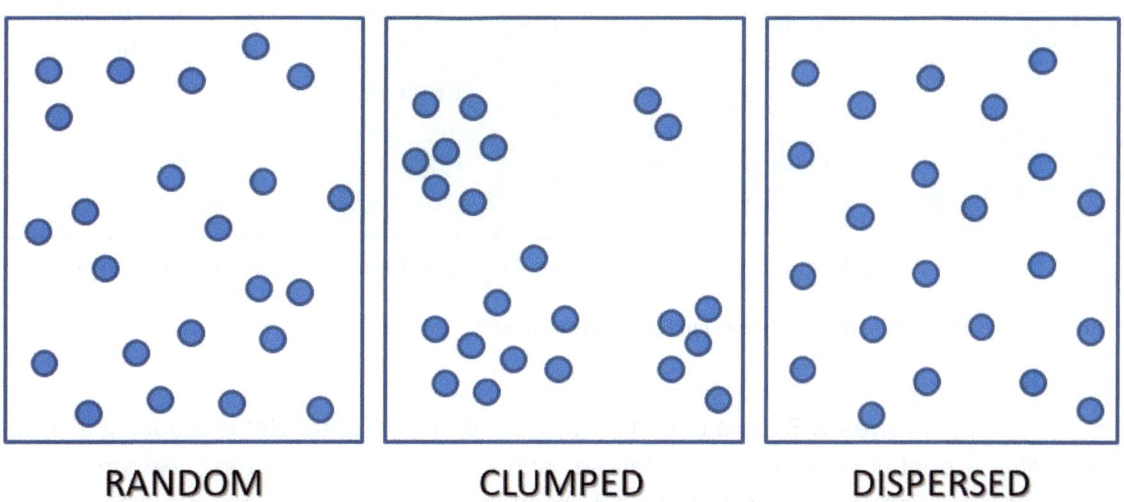

Georeferencing and digitizing

It is likely that at some point during your career, you will come across a photo or image that contains useful information, but is not spatially referenced. In other words, the photo or image is not in a format that is readily usable in a GIS. For instance, archived aerial photos may not be georeferenced. **Georeferencing** is the process of assigning coordinates to an image to identify its spatial location. Imagery collected from unmanned aerial vehicles or other aerial platforms are not initially georeferenced. In such cases, you may need to georeference the photos or imagery.

Often, georeferenced imagery is converted into a raster data model depicting land cover or it is digitized into points, lines, or polygons (i.e., a vector data model) depicting specific entities on the earth. For instance, aerial imagery may be used to digitize polygons representing ownership boundaries, roads, forest trails, study areas, or vegetation types.

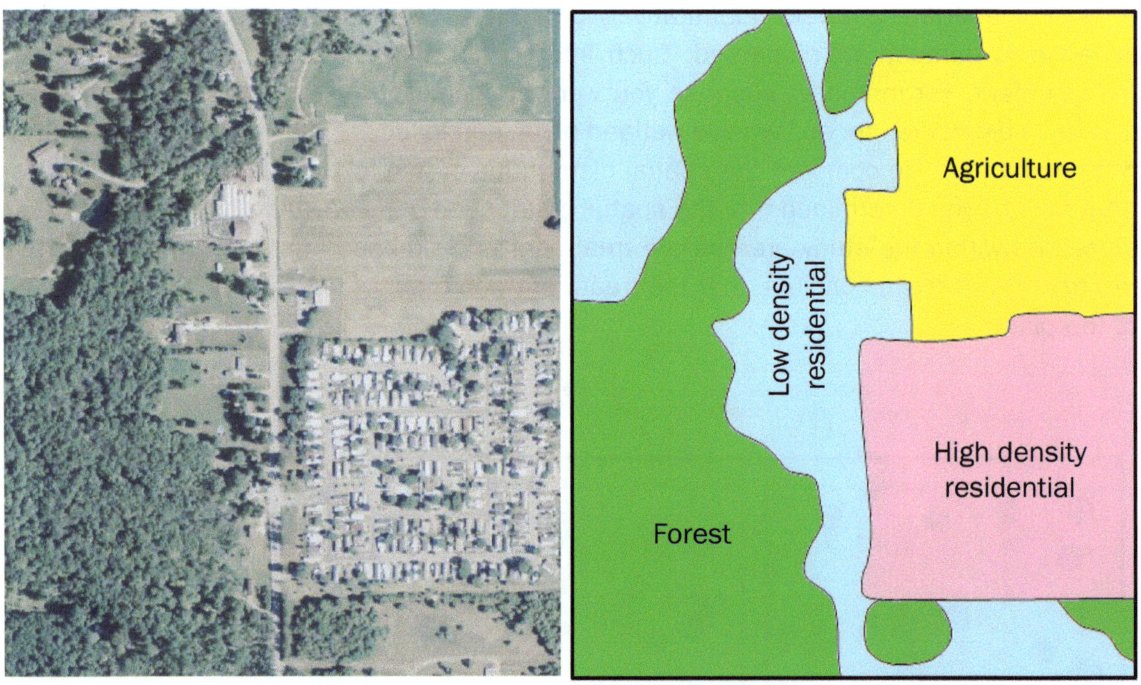

Aerial image   Digitized land cover (vector data model)

Source: National Agriculture Inventory Program, U.S. Department of Agriculture.

Geoprocessing

Geoprocessing involves GIS tools used to manipulate spatial data. Geoprocessing involves performing a spatial operation on a dataset and creating a new dataset. Geoprocessing may involve overlaying two datasets to clip or intersect them, establishing buffers around features, or interpolating point features to create continuous surfaces. There are numerous geoprocessing tools and techniques that can be used to manipulate and analyze spatial datasets. One who is an expert in GIS may not even be aware of all the geoprocessing tools available! Geoprocessing is used widely in spatial modeling and countless applications in natural resources management. Below is an example of simple geoprocessing where a linear feature such as a road is buffered and then intersected with land-cover data. One application of using this type of geoprocessing is to identify areas in which to test salt concentrations in the soil from road runoff during the winter.

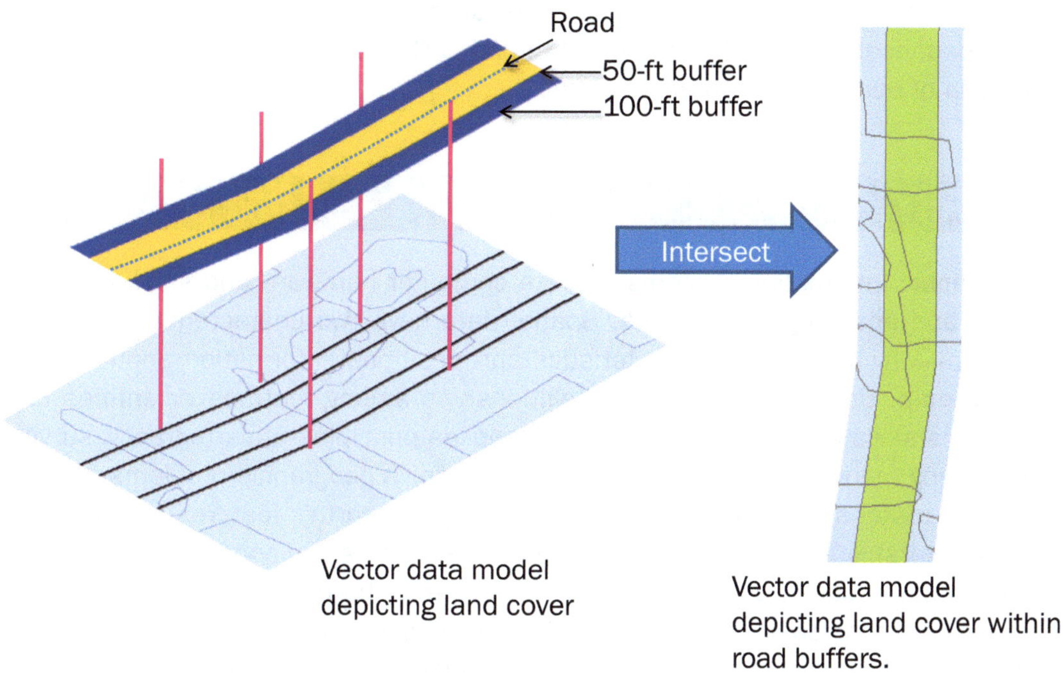

Applications of GIS Technology

GIS is widely used in many disciplines to better understand spatial relationships. For instance, GIS may be used to analyze patterns in the spread of disease, solve crimes, or determine the most profitable location for a new business. In natural resources-related fields, GIS skills are marketable and highly sought-after for their ability to help agencies and organizations better understand the earth, and the effects of management or disturbances on the earth's systems. For instance, land cover data may be used in a GIS to quantify the loss of vegetation types over time. We can quantify urban expansion rates, or areas at risk for erosion or pollution. We can use field data collected using GPS in conjunction with spatial data acquired through remote sensing to predict wildlife habitat suitability or identify areas in need of prescribed fire. We can map hiking trails, nesting box locations, urban landscapes, and watersheds with GIS. The applications are infinite! Advancements in GIS, GPS, and remote sensing technology facilitate the acquisition and availability of highly detailed spatial datasets. Many of these datasets are available for free through federal, state, county, or local agencies and organizations. For example, land cover between 2000 and the present day is publically available through the National Land Cover Database (Multi-Resolution Land Characteristics Consortium). Elevation, roads, archived historical imagery, waterbodies, tree canopy covers, and lots of other publically available datasets allow GIS users to analyze a plethora of spatial questions that were very difficult to address even a few short decades ago. Understanding the spatial relationships of our natural resources is critical for conservation and planning. GIS is, in part, a critical component of the future conservation of our earth's natural resources.

SUMMARY

GIS (Geographic Information Systems) ingrate computer hardware and software systems for viewing and analyzing spatial data. Spatial data have two characteristics—attributes and location—which describe characteristics and positional information about a specific entity on a map. Understanding spatial relationships among features or entities on the earth requires skills in spatial thinking, which applies principles of cartography, surveying, remote sensing, geography, math, and statistics to solve geographic problems. GIS maps use vector or raster spatial data models to represent the earth's features as points, lines, polygons (vectors), or pixels (rasters), respectively. GIS users can conduct spatial analyses and develop maps based on queries, pattern analyses, the creation of spatial data, and geoprocessing. An increasing amount of publicly accessible spatial data and advancements in spatial technology facilitate the analysis and understanding of spatial relationships on earth, and this has become progressively more important in natural resources conservation and planning.

TEST YOUR SKILLS

1. The data in the table on the next page represents point locations of fish caught on a lake. Suppose you were using the data to investigate spatial relationships between different species, their length, health, and location on the lake. Indicate whether each attribute is nominal, ordinal, interval, or ratio.

Species_code	Name	Length_inches	Condition	Date	Age_years
10	Largemouth bass	12	Healthy	16 Jun	5
11	Walleye	18	Healthy	16 Jun	4
11	Walleye	17	Healthy	17 Jun	4
20	Bluegill	5	Moderate	15 Jun	5
20	Bluegill	7	Healthy	17 Jun	6
21	Pumpkinseed	7	Poor	16 Jun	6
22	Perch	8	Poor	16 Jun	5

2. Suppose you are a wildlife biologist studying the relationship between habitat suitability and body mass of the brown-bellied hare (a fictitious species). You have created two maps: one map depicts average habitat suitability within 1-square-mile sections in your study area. The second map depicts average body mass of the hares based on individuals you trapped within 1-square-mile sections. Practice your spatial thinking skills. Use the two maps to infer the relationship between habitat suitability and body mass of brown-bellied hares. Describe the spatial distribution of habitat and hare size using spatial terms such as directions or descriptions of location.

Habitat suitability for brown-bellied hares. (Darker colors indicate better habitat).

Body mass of brown-bellied hares. (Darker colors indicate larger hares).

REFERENCES

Bolstad, P. (2008). *GIS Fundamentals: A First Text on Geographic Information Systems*, 3rd Edition. White Bear Lake, MN: Eider Press.

Goodchild, M. F. (1992). Geographical Information Science. *International Journal of Geographical Information Systems*, 6: 31–45.

Multi-Resolution Land Characteristics Consortium. (https://www.mrlc.gov/) Accessed March 15, 2019.

National Research Council. (2006). *Learning to Think Spatially*. Washington, DC: National Academies Press.

Zar, J. H. (1999). *Biostatistical Analysis*, 4th Edition. Upper Saddle River, NJ: Prentice Hall.

14 Vegetation Sampling with Plots

More than likely at some point in your career as an natural resources professional, you will be collecting data on vegetation structure and composition. **Vegetation structure** refers to the way in which vegetation is arranged. For instance, structural variables to describe a forest may include stem density, overstory canopy cover, tree diameter, shrub density, volume of down woody debris, or percent leaf litter. **Vegetation composition** refers to the type and relative amount of vegetation that occurs in the study area. For instance, you may describe the composition of a forest as 50% red maple, 30% aspen, and 20% balsam fir.

Vegetation structure and composition can be sampled using plots or plotless methods. The purpose of this lab is to provide you with experience conducting vegetation sampling using plots and analyzing data from plots.

LEARNING OBJECTIVES

1. Demonstrate how to use plots for sampling vegetation
2. Explain how nested plots are used for vegetation sampling
3. Define some common variables associated with vegetation structure
4. Describe how to measure common variables associated with vegetation composition
5. Calculate density and canopy cover using plots

Definitions

DBH: diameter at breast height; diameter of a tree measured at 4.5 ft or 1.37 m from ground
Density: number per unit area
Down woody debris: fallen deadwood
Canopy cover: % ground covered by the canopy of overstory trees
Ground cover: amount of plot covered by vegetation (i.e., not bare ground); measured as %
Midstory cover (shrub cover): % plot covered by shrubs
Snag: standing dead tree

Density of snags: count number of snag (standing dead stems)/200 m plot
Deciduous: perennial plants with leaves that shed (e.g., maples, oaks)
Coniferous: cone-bearing tree (e.g., pines, spruces)

Measuring/Calculating Variables

DBH

Using a diameter tape, measure around the bole of the tree 4.5 feet above the ground. Make sure to note what units (i.e., centimeters or inches) the tape uses. Tree diameter is important for characterizing the size of trees in a forested area. This information may be important for quantifying how much timber the forest can provide or for assessing wildlife habitat. Some types of wildlife prefer small trees, while others prefer forests with larger trees.

Density

Density is a measure of how crowded an area is with a particular feature such as trees, snags, rare plants, deer, or other wildlife. Count the total number of objects of interest within the plot. Then, divide by the plot size to get number/unit area. Convert area units if desired.

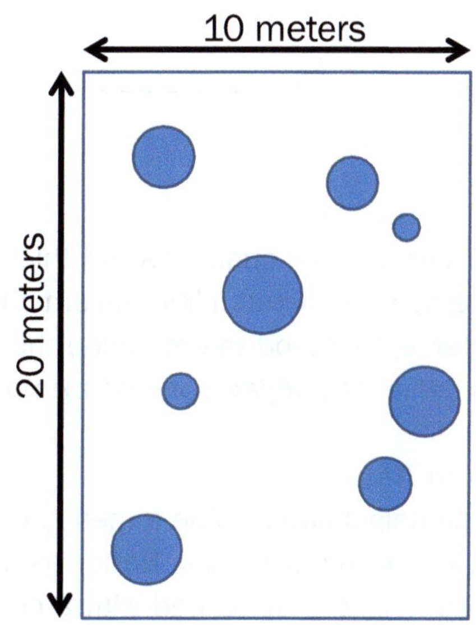

1. Count the number of trees

   $8 \; trees \,/\, plot$

2. Determine plot area

   $10 \times 20 = 200 \; m^2$

3. Divide number by area

   $$\frac{8 \; trees}{200 \; m^2} = 0.4 \; trees \,/\, m^2$$

4. Convert to meaningful area units such as number per hectare

   $$\frac{0.04 \; trees}{1 \; m^2} = \frac{10{,}000 \; m^2}{1 \; ha} = \mathbf{400 \; trees \; per \; hectare}$$

Canopy cover

Overstory canopy cover provides a metric to understand how open or closed the canopy is. Closed-canopy forests shade the ground, while open-canopy forests allow light to reach the forest floor. Use the line-intercept method to determine the proportion of a line that the canopy covers. Use one side of the plot length (20 m). Determine how much of the 20-meter line is covered by overstory vegetation.

Forest plot

Aerial view of plot.

Canopy cover is ~ 30% along the line intercept.

Midstory cover

Midstory cover quantifies the thickness of a forest. Midstory cover may be an important metric for understanding the amount of hiding cover for ground-nesting birds. Use the line-intercept method to determine the proportion of a line that the midstory covers. Midstory is defined as vegetation approximately 1–2 meters in height.

Tree height

Tree height can provide a measure of how tall a forest is or provide an index of the quality of a site to grow trees. Better sites grow taller trees at a given age than do poorer sites. Tree height is measured with a clinometer. Clinometers measure angles from the ground to the top of the tree. Stand 66 feet (1 chain length) from the base of an overstory tree in your plot. Have someone measure the height from the ground to your eye level. Keep both eyes open while you use a clinometer to read the angle from the ground to the top of the tree. Read the angle in degrees (the right side of the dial). Use the tangent of the angle to calculate tree height. Add it to your eye height to get the total tree height.

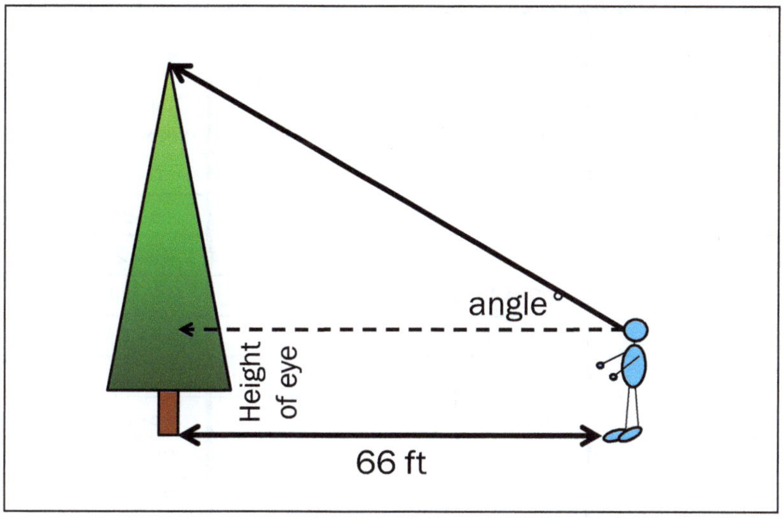

Example Data Sheet

Use the example data sheet on the following page to collect and summarize some vegetation structure data. Practice tabulating average tree diameter, densities, canopy cover, and tree height.

Plot 1			Plot 2			Plot 3		
Diameter class (cm)	Number of hardwoods	Number of conifers	Diameter class (cm)	Number of hardwoods	Number of conifers	Diameter class (cm)	Number of hardwoods	Number of conifers
0–4			0–4			0–4		
4–8			4–8			4–8		
8–12			8–12			8–12		
12–14			12–14			12–14		
14–16			14–16			14–16		
16–18			16–18			16–18		
18–20			18–20			18–20		
20–30			20–30			20–30		
30–50			30–50			30–50		
> 50			> 50			> 50		
Hardwood density			Hardwood density			Hardwood density		
Conifer density			Conifer density			Conifer density		
Stump density			Stump density			Stump density		
Log density			Log density			Log density		
Overstory cover			Overstory cover			Overstory cover		
Midstory cover			Midstory cover			Midstory cover		
Overstory height			Overstory height			Overstory height		

Averages across plots		
Diameter of hardwoods	Log density	
Diameter of conifers	Overstory cover	
Hardwood density	Midstory cover	
Conifer density	Overstory height	

SUMMARY

Natural resources professionals frequently collect vegetation data to understand how vegetation is arranged or composed within a landscape. Plots are often used to sample vegetation structure and composition within a study area. Common vegetation variables include tree diameter at breast height (DBH), the density of trees or shrubs, canopy covers, the density of snags or dead wood, and types of tree species present. Plots with a specific area and dimension are established randomly in a study site, and variables are typically recorded on a data sheet. Data are processed and summarized following collection.

TEST YOUR SKILLS

Suppose you work for the Forest Stewardship Program through the Department of Natural Resources. You have identified an invasive forest pest (emerald ash borer) in one of the state forests and you are charged with assessing the extent of the damage prior to developing a forest management plan. You want to determine how the forest structure will change after the ash trees have died. So, you acquired some high-resolution imagery of a portion of the forest (see below). You need to determine:

1. Density (#/acre) of ash trees
2. Density of non-ash trees
3. Percent canopy cover of ash trees
4. Percent canopy cover of non-ash trees

Procedure and critical information

Photo scale is 1:484.

You must create a plot using this photo scale. Cut a 2 × 2-inch window from a piece of paper. This window will serve as your plot.

Randomly select three locations on your photo. To ensure randomness, make a grid on the edges of your photo and mark each inch along the top and along the left side starting from the upper left (see diagram below). Each inch will be a coordinate. Randomly select three coordinates and use these coordinates as the upper left corner of three plots.

Sample ash snag density, density of non-ash trees, canopy cover of ash, and canopy cover of non-ash trees. Also, using the scale, measure the average width of future canopy openings as the ash snags fall to the ground.

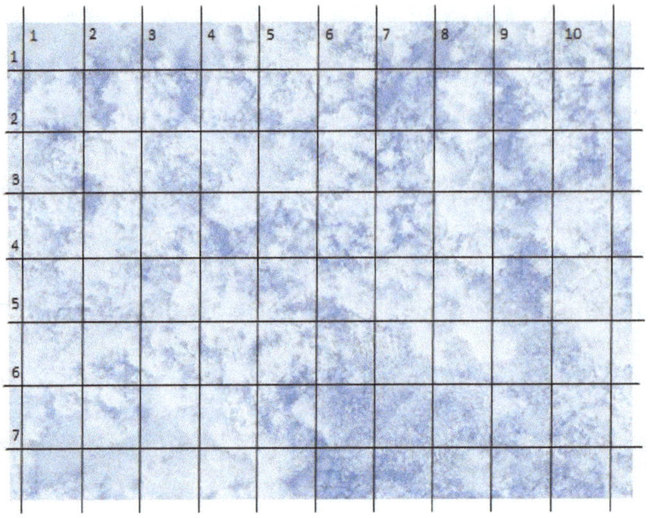

Suppose a randomly selected set of co-ordinates was 3, 4. The plot would be positioned within the bold region.

Practice summarizing your data in a scientific report with the following components:

Introduction: Describe the purpose of the study and the objectives.
Study area: Describe the location, climate, soils, vegetation.
Methods: Explain the procedure you took to conduct the study. Be sure to explain all variables measured.
Results: Calculate the ash density, density of non-ash trees, canopy cover of ash, and canopy cover of non-ash trees. Type your results in Tables 14-1 and 14-2. Describe ALL the data in tables 1 and 2, also typed, using complete sentences, as you would in a scientific report. Reference each table in the text as you describe it.
Discussion: Describe how the structure of the forest may change once the ash snags have fallen. What might be the implications of this change on forest composition, wildlife habitat, nutrient cycling, hydrology, and/or aesthetics?

Table 14-1 Measurements from aerial photo

	Number of ash snags	Number of non-ash trees	Length of plot intercepted by ash (in)	Length of plot intercepted by non-ash (in)
Plot 1				
Plot 2				
Plot 3				

Table 14-2 Density and percent cover of ash and non-ash trees and average width of openings created by fallen ash trees affected by emerald ash borer

	Density of ash snags (#/ac)	Density of non-ash trees (#/ac)	% canopy cover ash	% canopy cover non-ash	Average width of openings created by fallen ash (ft)
Plot 1					
Plot 2					
Plot 3					

REFERENCE

Avery, T. E. & Burkhart, H. E. (2002). *Forest Measurements*, 5th Edition. New York, NY: McGraw-Hill.

15

Vegetation Sampling Using Plotless Methods

Plots are great for sampling distributions and densities of populations. However, the results of plot data analyses often depend on the size and shape of the plot and the area under study. For instance, the distribution of individuals within the population under study may be better investigated with long linear plots versus square plots. Or perhaps circular plots would be better than rectangular plots. Also at times, plot set-up may be time consuming. So, in some instances, sampling vegetation or sessile organisms may be better with plotless methods.

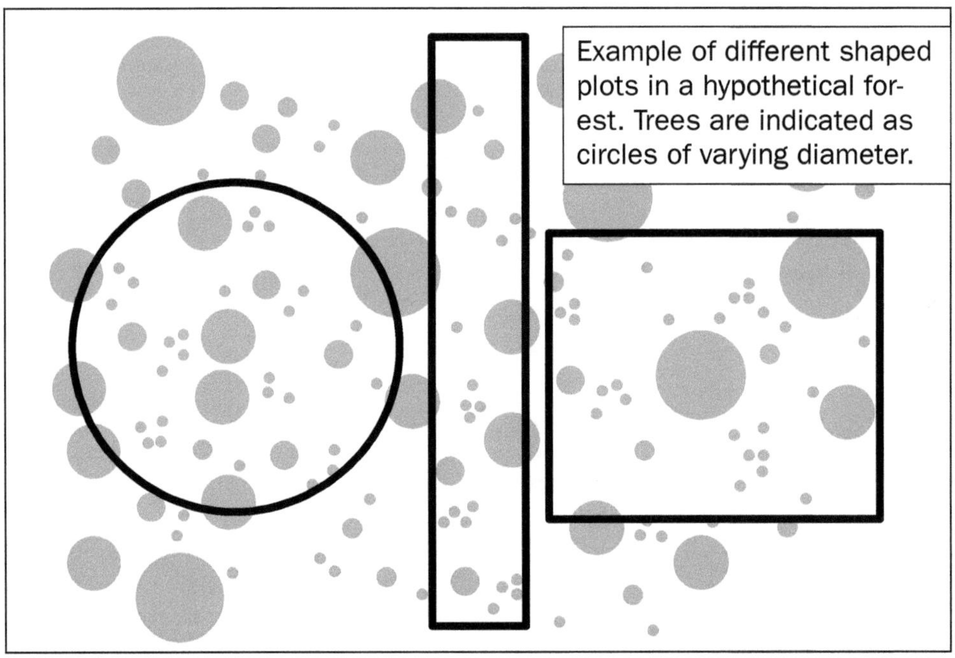

Example of different shaped plots in a hypothetical forest. Trees are indicated as circles of varying diameter.

Methods that do not use plots are often faster for sampling plants and sessile animals, and they do not depend on shape of sampling area. The point-quarter method is a common plotless method for sampling vegetation. This method can be used effectively in areas where vegetation is randomly distributed. Uniform or clumped distributions will yield biased results.

The purpose of this lab is to provide you with experience conducting vegetation sampling using the point-quarter method and analyzing data collected from this method.

LEARNING OBJECTIVES

1. Understand how to measure variables associated with vegetation structure using the point-quarter method
2. Distinguish between absolute and relative metrics
3. Calculate density, frequency, coverage, and importance values

Definitions

Absolute: the actual measurement of something
Relative: a metric considered in relation to or in proportion to something
Basal area: the area occupied by tree stems
Density: number per unit area
Frequency: the rate at which something occurs
Coverage: amount of area covered by somethign (e.g., vegetation)
Importance value: an index to assess the dominance or prevalence of a species in a community; higher importance values equate to a stronger presence

Point-Quarter Method

This sampling method is quick! Walk a straight-line transect on a set azimuth until you reach your point (say, 100 ft from your starting location). At your sampling point, visually establish four quadrants. Essentially, visualize a line perpendicular to your transect. In each quadrant, measure the distance to the closest object of interest. In this case, we are sampling trees. But you could be measuring snags, shrubs, or herbaceous vegetation.

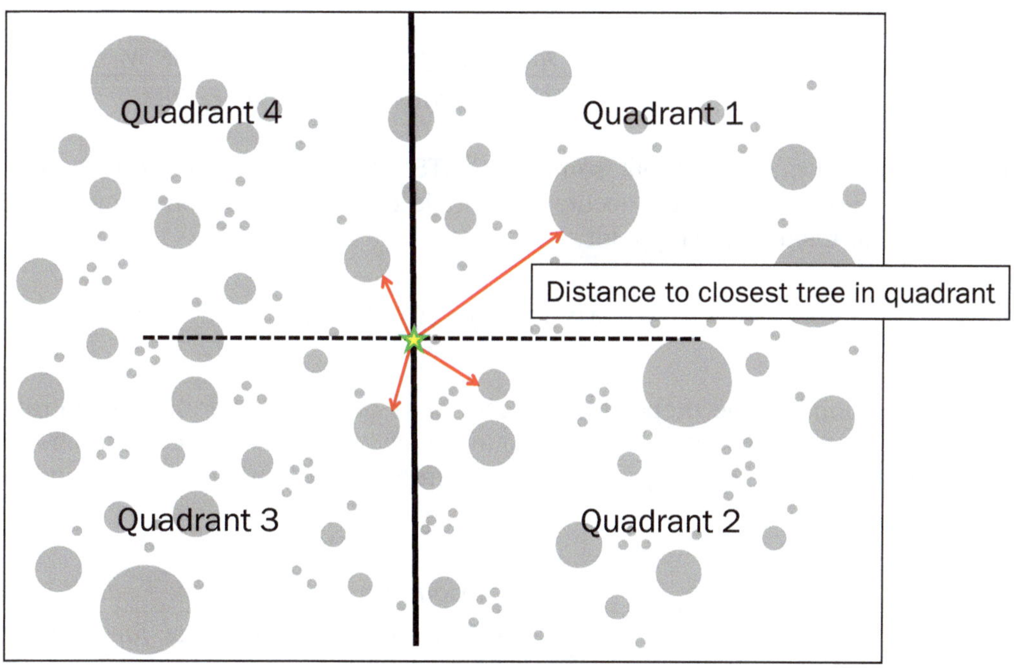

Because we are measuring trees, also measure the DBH. Once you have measured the distance and diameter to the closest tree in each quadrant, continue pacing 100 ft along your transect. At that point, repeat the sampling procedure.

Processing Point-Quarter Data

Suppose you work for an environmental consulting company that is contracted by a landowner who wants to understand his forest structure. You implemented the point-quarter method to quickly inventory the overstory trees in a 30-ha central hardwood forest (SM = sugar maple, RO = red oak, BW = basswood, WA = white ash). Below are point-quarter data you collected. DBH was recorded in cm. Distance was recorded in m. In the table on page 6, summarize the following: frequency, density, basal area, dominance, and importance for each overstory species. Then, describe the data, using complete sentences as you would in a scientific report.

	Quad 1			Quad 2			Quad 3			Quad 4		
Point	**spp**	**dbh**	**distance**	**spp**	**dbh**	**distance**	**spp**	**dbh**	**distance**	**spp**	**dbh**	**distance**
1	SM	33	5.2	RO	32	2.2	RO	34	6.1	SM	33	5.6
2	RO	34	5.6	SM	33	5.8	SM	34	5.4	BW	27	6.8
3	BW	27	6.7	RO	31	4.7	RO	35	2.4	RO	36	7.1
4	RO	43	2.3	BW	26	6.5	WA	28	5.3	RO	40	5.2

Absolute frequency = # of sampling points (NOT QUADRANTS) a species was observed/ total number of samples

$$AF\ for\ red\ oak = \frac{number\ of\ sampling\ points\ where\ red\ oak\ was\ observed}{total\ number\ of\ sample\ points}$$

e.g., Red oak: occurs at all four sample POINTS. Thus, the frequency of red oak is 4/4 = 1.0. It does not matter that it occurs in multiple quadrants, as frequency is a measure of occurrence at each sampling point.

$$AF\ for\ red\ oak = \frac{4}{4} = 1.0$$

Relative frequency = the proportion of occurrence of a particular species in relation to all other species. To calculate relative frequency, divide absolute frequency by the sum of absolute frequencies for all species. Relative frequency will be a proportion (i.e., a number between 0 and 1).

$$RF\ for\ red\ oak = \frac{absolute\ frequency\ for\ red\ oak}{the\ sum\ of\ all\ absolute\ frequencies\ for\ all\ species}$$

Total stem density = stems per area for all species combined. To calculate total stem density, first calculate the mean distance from sampling points to all trees. Next, calculate mean area by squaring mean distance. Finally, convert to appropriate units (e.g., hectares).

$$Mean\ distance = 1\ stem\ per\ 5.18\ m$$

$$Mean\ area = 1\ stem\ per\ (5.18\ m)^2 = 26.84\ m^2$$

$$Total\ stem\ density = \frac{1\ stem}{26.84\ m^2} * \frac{10{,}000\ m^2}{ha} = 372.6\ \boldsymbol{stems / ha}$$

Relative stem density = the proportional density of a particular species in relation to all other species. To calculate relative stem density, divide the number of stems of a species by the total number of stems counted.

$$RD\ for\ red\ oak = \frac{number\ of\ red\ oak\ stems}{total\ number\ of\ stems}$$

$$RD\ for\ red\ oak = \frac{8}{16} = 0.5$$

Absolute stem density = the actual stem density per species. Absolute stem density for a particular species is the relative stem density multiplied by the total stem density.

$$AD\ for\ red\ oak = relative\ density\ for\ red\ oak * total\ stem\ density$$

$$AD\ for\ red\ oak = 0.5 * 372.6\ stems / ha = 186.2\ \boldsymbol{red\ oak\ stems / ha}$$

Basal area = the area occupied by tree stems. Basal area is calculated using diameter at breast height. Basal area can be reported as the average area occupied per tree (e.g., m^2) or the area occupied by all trees within a specified area (m^2/ha).

$$BA\ per\ tree = \frac{\pi * dbh^2}{4} \quad [units\ will\ be\ in\ cm^2]$$

To calculate mean basal area per tree species, you must first calculate basal area per tree and then take the average. For example, the mean basal area for red oak is calculated as follows:

Quad	Tree	DBH (cm)	Basal area
Quad 1	RO	34	$(\pi * 34^2) / 4 = 907.5\ cm^2$
	RO	43	$(\pi * 43^2) / 4 = 1{,}451.5\ cm^2$
Quad 2	RO	32	$(\pi * 32^2) / 4 = 803.8\ cm^2$
	RO	31	$(\pi * 31^2) / 4 = 754.4\ cm^2$
Quad 3	RO	34	$(\pi * 34^2) / 4 = 907.5\ cm^2$
	RO	35	$(\pi * 35^2) / 4 = 961.6\ cm^2$
Quad 4	RO	36	$(\pi * 36^2) / 4 = 1{,}017.4\ cm^2$
	RO	40	$(\pi * 40^2) / 4 = 1{,}256\ cm^2$

Mean basal area for red oak can be calculated as follows:

$$Mean\ BA\ for\ red\ oak = \mathbf{1007.4}\ \boldsymbol{cm^2}$$

Absolute dominance = a measure of the amount of area occupied by a species. This is reported in terms of basal area on a per-hectare or per-acre basis. The units are usually m^2/ha or ft^2/ac. So, you may have to convert the mean basal area from cm^2 to m^2 or from in^2 to ft^2.

$$BA\ per\ hectare = mean\ BA\ per\ species * stem\ density\ of\ the\ species$$

$$BA\ of\ red\ oak\ per\ hectare = \frac{1{,}007.4\ cm^2}{1} * \frac{1\ m^2}{10{,}000\ cm^2} * \frac{186.2\ stems}{ha}$$

$$BA\ of\ red\ oak\ per\ hectare = \mathbf{18.8}\ \boldsymbol{m^2 / ha}$$

Relative dominance = the proportional dominance of a particular species in relation to all other species. To calculate relative dominance, divide the BA per hectare of species by the total basal area of all species combined.

$$Relative\ Dominance\ for\ red\ oak = \frac{BA\ of\ red\ oak\ per\ hectare}{total\ BA\ for\ all\ species\ combined}$$

Importance value = an index of the relative contribution of different species to the composition of the forest. The importance value of a species is calculated by adding relative frequency, relative density, and relative dominance.

$$IV\ for\ red\ oak = relative\ frequency + relative\ density + relative\ dominance$$

SUMMARY

In some instances, vegetation sampling may occur with plotless sampling methods. The point-quarter method is a common sampling method in areas where vegetation is randomly distributed. In the point-quarter method, an observer walks a specific distance along a transect. At the sampling point, the observer visualizes a line perpendicular to the transect, which establishes four quadrants. The distance from the sample point to the closest object in each quadrant is measured. Other characteristics of the closest object (e.g., tree) may also be recorded, but they are only recorded from the closest object to the sample point. Metrics including absolute and relative frequency, density, and dominance may be computed from point-quarter vegetation data.

TEST YOUR SKILLS

1. Complete the table below with calculations for the other tree species in the data provided above.

2. Use the sample study area and data sheet below to collect vegetation data using the point-quarter method. Compute absolute and relative frequencies, densities, dominance, and importance value.

Table 15-1 Summary of vegetation composition from point-quarter data

Species	Absolute frequency	Relative frequency	Total number of trees	Relative density	Absolute density	Mean basal area per tree (cm^2)	Absolute dominance	Relative dominance	Importance value
red oak	1.0		8	0.5	186.2	1,007.4	18.8		
sugar maple									
white ash									
bass-wood									
TOTALS		1.0	16	1.0	372.6			1.0	

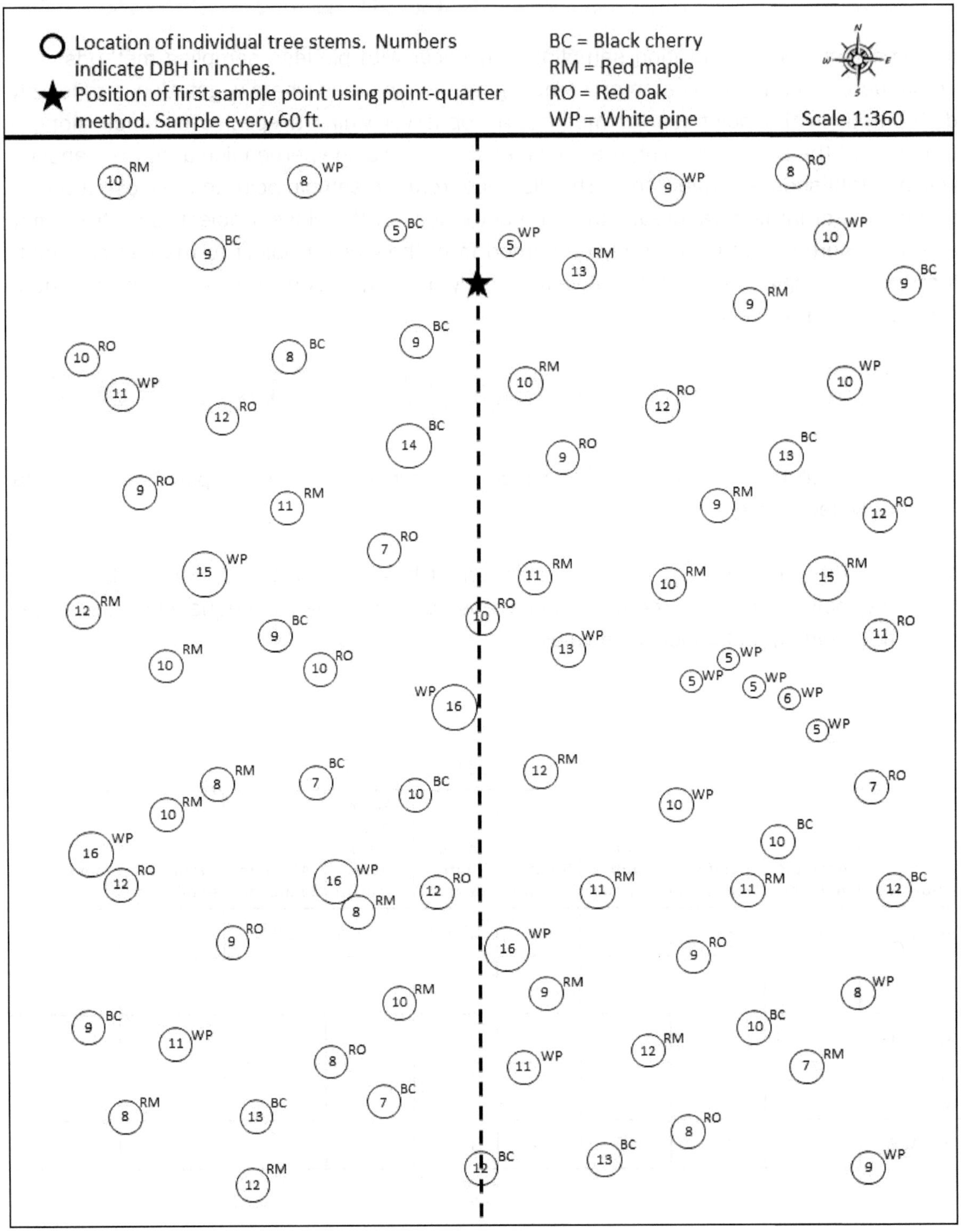
Location of individual tree stems. Numbers indicate DBH in inches.
Position of first sample point using point-quarter method. Sample every 60 ft.
BC = Black cherry
RM = Red maple
RO = Red oak
WP = White pine
N
W
E
S
Scale 1:360

Sample study area used for sampling trees using the point-quarter method
Vegetation sampling datasheet using the point-quarter method

Group Members: ____

Point No.	Quad No.	Species	Diameter (cm)	Distance from point (m)
1	1			
	2			
	3			
	4			
2	1			
	2			
	3			
	4			
3	1			
	2			
	3			
	4			

Table 15-2 Summary of vegetation composition using sample data and the point-quarter method

Species	Absolute frequency	Relative frequency	Total number of trees	Relative density	Absolute density	Mean basal area per tree (cm^2)	Absolute dominance	Relative dominance	Importance value
black cherry									
red maple									
red oak									
white pine									
TOTALS		1.0	12	1.0				1.0	

REFERENCE

Avery, T. E. & Burkhart, H. E. 2002. *Forest Measurements*, 5th Edition. New York, NY: McGraw-Hill.

16

Descriptive Statistics

Data analysis is imperative in any natural resources–related job. As resource managers, we constantly collect data that need to be analyzed for status reports, planning, adaptive management, policy development, grant acquisition, and many other reasons. **Statistics** are ways of summarizing and describing quantitative data to draw inferences on the data and understand whether results are characteristic of a population or due to random chance. Many statistical tests have been developed for various types of data and types of questions. Any researcher should have a solid foundation in the field of statistics. Other professionals should, at minimum, be able to summarize data using descriptive statistics. **Descriptive statistics** describe the main features in a data sample. They allow an understanding of population tendencies and variability. For instance, a mean and standard deviation are common metrics describing the average and variability within a sample. This chapter is intended to provide a fundamental understanding of basic sampling concepts and descriptive statistics.

LEARNING OBJECTIVES

1. Distinguish between biological population and statistical population
2. Define terms associated with basic sampling
3. Define *mean*, *median*, *mode*
4. Define metrics associated with population variability
5. Compute basic descriptive statistics from sample data

Sampling Concepts

Suppose you are a fisheries biologist comparing the length of northern pike between lakes with high and low fishing pressure. Based on preliminary observations and chatter among anglers, you hypothesize that pike are smaller on lakes with high fishing pressure because anglers tend to keep the larger ones. To support establishment of fishing regulations, you collect some evidence to illustrate the size difference in fish between lakes with high and low fishing pressure. In this report, the quantitative data must be summarized using basic descriptive statistics. But first, an understanding of sampling concepts is critical.

Population: biological versus statistical

The two lakes you are investigating are Bear Lake and Otter Lake. All of the northern pike in Bear Lake can potentially interact and breed with one another. In essence, all the pike in Bear Lake are members of the same **biological population**. Otter Lake has two parts: a northern area and a southern area separated by a sandbar. The pike in the northern area do not interact with those in the southern area because the sandbar acts as a barrier to breeding. Thus, there are two biological populations in Otter Lake. A statistical population is different. A **statistical population** is one that is the subject of investigation. In this case, one statistical population consists of all of the pike in Bear Lake and another statistical population consists of all the pike in Otter Lake. Essentially, you are collecting data to make inferences about the pike populations in each lake. In some instances, a statistical population may be composed of several biological populations, or alternatively, a statistical population may be composed of individuals representing part of a biological population.

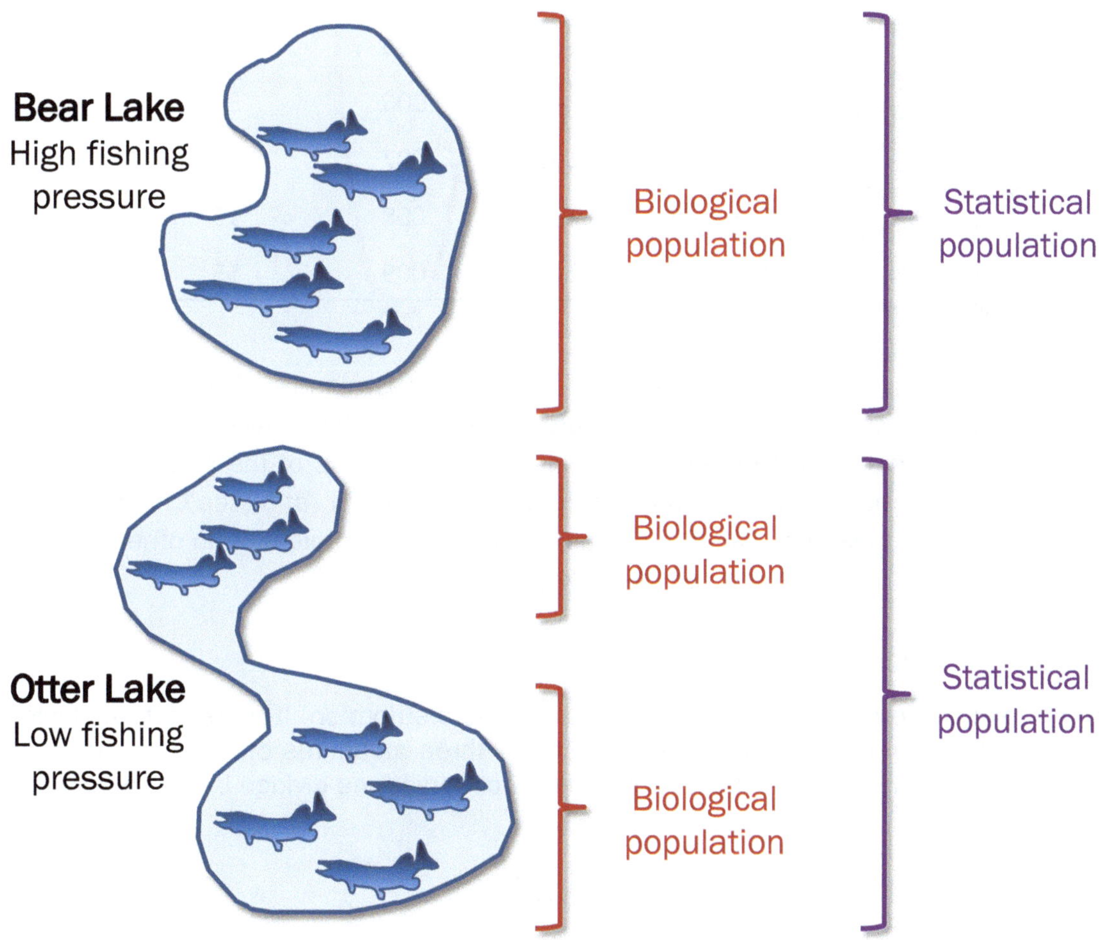

Sampling

To determine whether or not pike in Bear Lake are smaller than those in Otter Lake due to differences in fishing pressure, you must measure fish in each lake. However, it is completely unreasonable and impractical to measure all pike in both lakes. So, you collect a sample. A **sample** is all of the individuals from a statistical population that are measured for the study. Thus, you will have two samples: one from Bear Lake and one from Otter Lake. The samples collected must be **random**; that is, they must be free from bias. In a random sample, every individual must have an equal probability of being caught. So, you randomly place hoop nets throughout each lake to capture and measure your samples. A **sampling unit** is the unit on which a measurement is taken. Because you are measuring length of pike, the sampling unit in this case is an individual fish. Length, in this case, is a **variable**. Variables are attributes that differ between individuals in a population and what is being measured to infer characteristics of the population. Recall that variables can be classified as nominal, ordinal, interval, or ratio (see Chapter 13). An **observation** is the measured value or characteristic of one sample unit. In other words, an observation is the length of one sampled pike.

Sample

Sample unit

Variable

Observation

Lake	Date	Pike ID	Length (cm)
Bear Lake	11-Jul	001	45
Bear Lake	11-Jul	002	20
Bear Lake	11-Jul	003	10
Bear Lake	11-Jul	004	34
Bear Lake	11-Jul	005	20
Bear Lake	12-Ju	006	30
Bear Lake	12-Jul	007	68
Bear Lake	12-Jul	008	15
Bear Lake	12-Jul	009	55

Measures of Average

The average is the central tendency within a sample distribution. It represents a "typical" value within a sample dataset. There are actually three measures of average: mean, median, and mode. The **mean** is the sum of a set of observations divided by the total number of observations.

$$\bar{x} = \frac{\sum x}{n}$$

Where $\overline{x}$ is the sample mean, x is the value of the observation, and n is the sample size (i.e., the total number of observations you have). For example, the mean of the length of the nine pike sampled in Bear Lake (above) is 33.0 cm.

$$\overline{x} = \frac{(45+20+10+34+20+30+68+15+55)}{9}$$

$$\overline{x} = 33.0\ cm$$

The **median** is the middle observation in a set that has been ranked in magnitude. So, if we ranked the observations of pike length from low to high, the middle value would be 30.

10 15 20 20 **30** 34 45 55 68

The **mode** is the most frequently occurring value in a sample. For example, in the pike sample dataset, the most frequently occurring value is 20. There are two pike measuring 20 cm. There is only one pike with each of the other observed lengths. A distribution may be unimodal (i.e., having one mode), bimodal (i.e., having two modes), or multimodal.

The mean, median, and mode all represent a location within a distribution. A **sample distribution** shows the frequency of occurrence of observations within a sample. For example, suppose you actually sampled 100 pike from each lake and plotted the sample distribution. The distribution of pike lengths in Otter Lake is normal. That is, it looks like a bell-shaped curve with equivalent values for the mean, median, and mode. On the other hand, the sample distribution for Bear Lake is skewed. That is, there is a larger number of fish with smaller lengths than fish with larger lengths. Thus, the mean, median, and mode occur at different locations on the distribution.

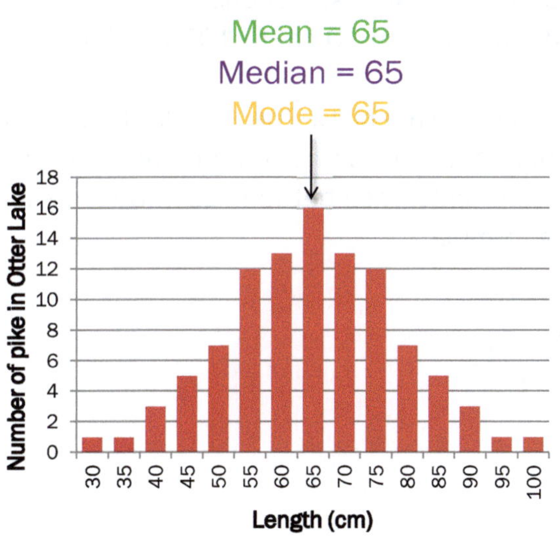

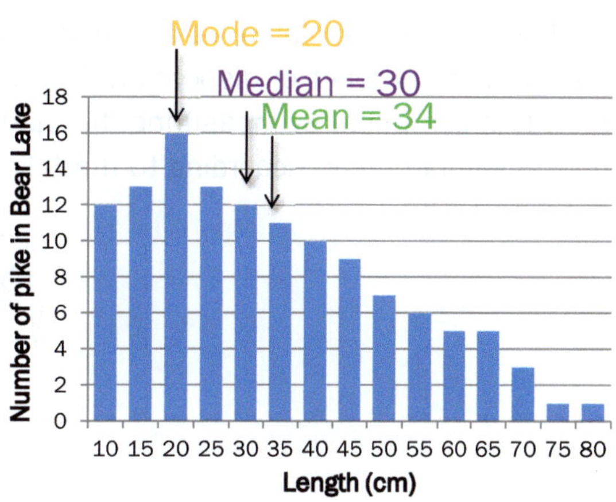

Measures of Variability

Measures of average are not quite enough to describe the general characteristics of a dataset. Measures of **variability** describe the extent to which data points differ from each other. There are three common measures of variability: range, standard deviation, and variance. Sample distributions with lower variability are narrower than those with higher variability.

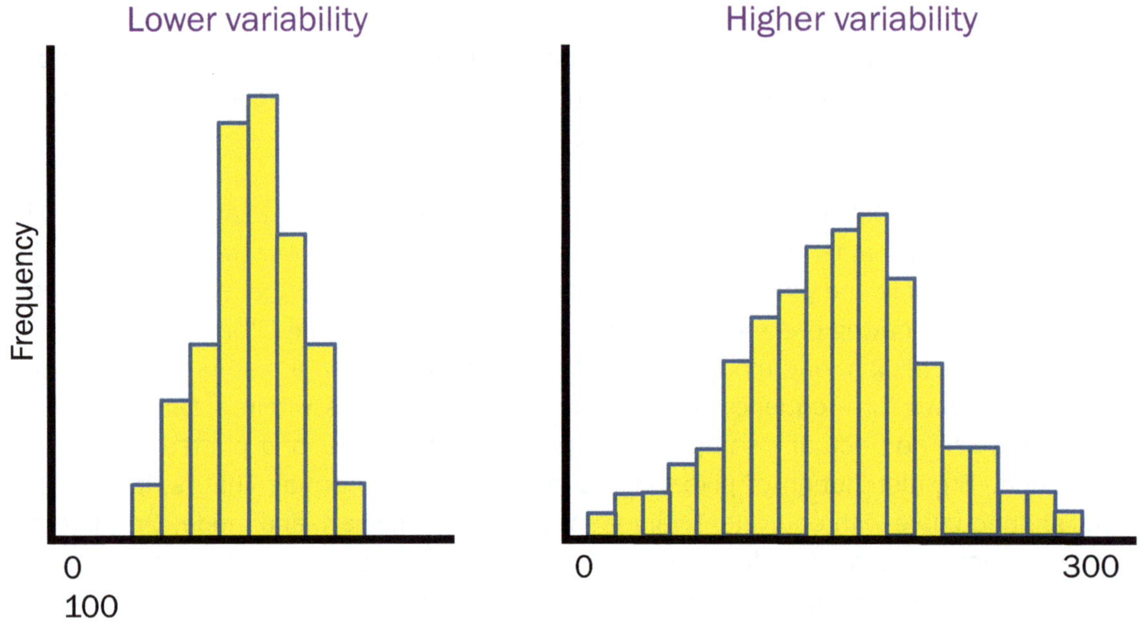

The **range** is the magnitude between the highest and lowest observations. For instance, in the 100-pike sample from Otter Lake, the lengths range from 30 cm to 100 cm. Thus, the range is 70 cm.

The **standard deviation** is a common measure of variability. The standard deviation measures the amount of dispersion that a sample dataset has from the mean of the dataset. It is computed by measuring the magnitude of difference between each observation and the sample mean according to the following formula:

$$s = \sqrt{\frac{\sum (x - \bar{x})^2}{n-1}}$$

Where s = an estimate of the standard deviation, x = an observation, $\bar{x}$ = the sample mean, and n = the sample size. Let's compute the standard deviation for the subsample of nine observations from Bear Lake (indicated above).

Step 1. Compute the difference between each observation and the sample mean $(x - \bar{x})$.

Step 2. Square the difference between observation and the sample mean $[(x - \bar{x})^2]$.

Step 3. Sum (Σ) all the values you computed in Step 2. This is referred to as the sum of squares.

Observations Step 1 Step 2

Pike ID	x	$x - \bar{x}$	$(x - \bar{x})^2$
001	45	12	144
002	10	-23	529
003	15	-18	324
004	20	-13	169
005	20	-13	169
006	30	-3	9
007	34	1	1
008	55	22	484
009	68	35	1225

Σ =3054 Step 3

Step 4. Divide the value you computed in step 3 by the sample size (n = 9) minus 1 (9 – 1 = 8).

Step 5. Take the square root of the value you computed in step 4.

$$s = \sqrt{\frac{3{,}054}{8}} = 19.5\ \boldsymbol{cm}$$

The standard deviation of this subsample of nine observations from Bear Lake is 19.5 cm. This value means that characteristically, the sample observations are concentrated within 19.5 cm of the mean. Some observations deviate a greater distance from the mean, and some do not deviate so much. The empirical rule of thumb states that in a normally distributed sample, nearly 100% of the observations in the distribution lie within three standard deviations of the mean. If we graphed the observations of pike length from Bear Lake in relation to the mean, we can visualize the standard deviation as follows:

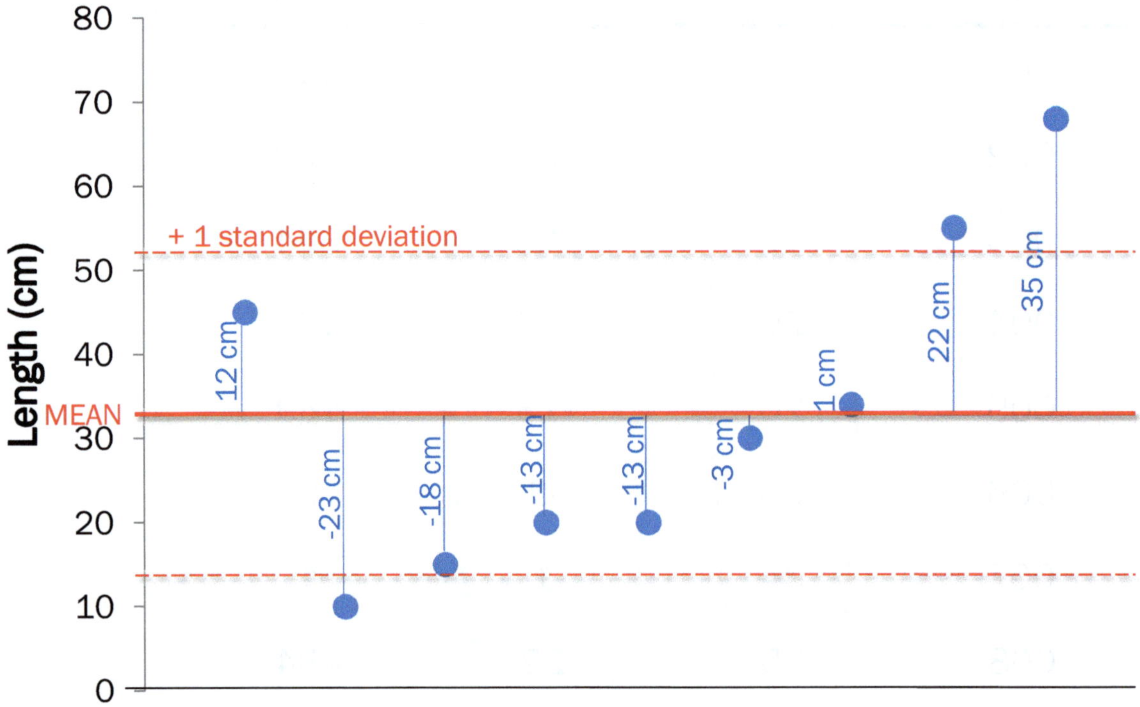

The final measure of variability is the **variance**. The variance is the standard deviation squared. Alternatively, the standard deviation is the square root of the variance. The variance is used in some statistical evaluations you may learn about in a statistics course.

$$s^2 = \frac{\sum (x - \bar{x})^2}{n-1}$$

Measuring Error

The **standard error** of a mean quantifies how precisely you have estimated the true population mean. The standard error tends to be larger with small sample sizes than with large sample sizes. In other words, the sample mean is probably a better estimate of the true population mean when the sample size is larger. The standard error is computed by dividing the standard deviation by the square root of the sample size.

$$S.E. = \frac{s}{\sqrt{n}}$$

The standard error is used to compute a **confidence interval**. Confidence intervals indicate a range of values within which the true population mean is likely to occur, given the sample data. Wider confidence intervals for a particular sample are less precise than smaller ones. Typically, a 95% confidence limit is reported for sample means. A 95% confidence interval indicates the range of values in which the true population mean would occur 95% of the time with repeated sampling. In samples containing a relatively large number (e.g., n = 30) of observations, we are 95% confident that the true population mean falls within the sample mean ± 1.96 * S.E.

$$95\%\ confidence\ interval = \overline{x} \pm 1.96 * S.E.$$

Small samples (e.g., n < 30 observations) have greater uncertainty than larger samples. Therefore, confidence intervals for means of small samples must apply a correction factor. The number 1.96 in the formula above is replaced with a suitable correction determined by the sample size and statistical tables such as a t-table. These procedures are taught in basic statistics courses.

SUMMARY

Data analysis and summarization is imperative in any natural resources-related job. At minimum, basic descriptive statistics are essential for summarizing the main features in a dataset. They allow an understanding of population tendencies and variability. To make inferences about a statistical population, a researcher would collect a random sample of individuals and record observations about each one. Descriptive statistics would report the central tendencies of the observations or measurements through the mean, median, or mode. Measures of variability include the standard deviation, range, and variance. Error can be estimated using metrics such as the standard error of a mean and the confidence interval.

TEST YOUR SKILLS

Suppose you were conducting a habitat analysis to investigate the selection of nesting locations of the blue-footed whistlebird and the red-bellied whistlebird. One of the variables you are investigating is distance to water. So, you plot the GPS coordinates of known nest locations on a map and measure the distance from them to the nearest open water source for each bird species. Your objectives are to 1) report descriptive statistics associated with nesting locations of each bird species and 2) recommend a distance to water at which nest boxes should be placed to increase nesting habitat for each bird species.

Please analyze the data and answer the questions below as part of your assessment of nest locations.

Table 16-1 Distances in meters of nesting locations to open water for two bird species

	blue-footed whistlebird	red-bellied whistlebird
1	304	242
2	259	233
3	298	240
4	245	242
5	260	235
6	245	256
7	265	238
8	248	243
9	241	242
10	253	234

1. What is the sampling variable?
2. What type of variable is it (nominal, ordinal, interval, ratio)?
3. Report the mean distance to water for each bird species? Report the median and mode.
4. What is the range of distances for each bird species?
5. What is the standard deviation of distance to water for each bird species?

6. Compute the 95% confidence interval of nest distance to water for each bird species. Note: because the sample size is < 30 observations, the confidence interval must be computed with the following formula:

$$95\%\ \textit{confidence interval} = \bar{x} \pm 2.262 * S.E.$$

7. Based on the confidence intervals, at what distance to water should you place nesting boxes for each bird species to increase their use? Should the recommended distance be the same for both species? Provide evidence for your answer.

REFERENCES

Fowler, J., Cohen, L, & Jarvis, P. (1998). *Practical Statistics for Field Biology*, second edition. Chichester, England: John Wiley & Sons.

Heath, D. (1995). *An Introduction to Experimental Design and Statistics for Biology*. London, England: UCL Press Limited.

Zar, J. H. (1999). *Biostatistical Analysis*, Fourth Edition. Upper Saddle River, NJ: Prentice Hall.

17

Glossary

Absolute: the actual measurement of something

Absorbed energy: energy that is converted into some other form, such as light being converted into heat

Active remote sensing: the acquisition of information by generating an energy signal and recording the energy returned (e.g., Doppler radar)

Aerial image: a digital image of the earth's surface taken from a point above the surface

Aerial photo: a hard copy picture of the earth's surface taken from a point above the surface

Agonic line: areas on the earth where declination is 0°

Analemma: a figure 8 diagram showing the position of the sun as seen from a fixed location on the earth

Association: the tendency of some objects to occur together or within close proximity

Astrolabe: an instrument used to measure the positions of celestial bodies; historically used to calculate latitude before the sextant

Attributes: the characteristics of a specific entity

Azimuth: an angular measurement between 0–360° on a sphere

Basal area: the area occupied by tree stems

Base line: in the US Public Land Survey System, the east–west line used for survey reference; divides townships between north and south

Bearing: a land navigation measurement referring to the angle that describes position in relation to a north–south line

Benchmark: a position with a known elevation of one, from which topographic survey measurements will be referenced

Biological population: all organisms within a species that can potentially interact and breed with one another

Boundary survey: the establishment of the corners of a parcel of land, and the identification of its location using a legal description

Canopy cover: the percent of the ground covered by the canopy of overstory trees in a forest

Cartesian coordinate system: a coordinate system that defines position using right-angle x- and y-axes (or z-axis if elevation is included); created through map projections

Chronometer: an extremely accurate clock initially designed to determine longitude for marine navigation

Color: in photogrammetry, the variations in the reflectivity of different objects

Compass: an instrument used for navigation that shows the four cardinal directions

Compass baseplate: the structure on which a compass is mounted

Compass degree dial: on an orienteering compass, a rotating feature that houses the magnetic arrow

Compass index line: the mark on a compass that indicates where the azimuth or bearing should be read

Confidence interval: a range of values within which the true population mean is likely to occur

Conifer: a cone-bearing tree (e.g., pines, spruces)

Contour interval: the vertical distance between contour lines on a map

Contour lines: lines on a map that connect points of equal elevation

Contour map: a surface map that uses lines to depict the shape and elevation of land

Convergence of evidence: in photogrammetry, a process whereby several elements (i.e., shape, shadow, tone/color, texture, pattern, site, and association) are considered collectively in order to determine the proper classification or identification of objects

Coordinates: a set of values that define location in two- or three-dimensional space

Cosine: the trigonometric function of an angle in a right triangle equal to the length of the side adjacent to the angle divided by the length of the hypotenuse

Coverage: the amount of area covered by something (e.g., vegetation)

DBH: diameter at breast height; refers to the height at which the diameter of trees is typically measured

Deciduous: perennial plants with leaves that shed in autumn (e.g., maples, oaks)

Declination: the difference between magnetic north and true north

Density: the number per unit area

Descriptive statistics: the main features in a data sample

Differential leveling: the process of collecting elevation data along line transects or within a specific area of interest

Direction of travel arrow: the arrow on the baseplate of a compass that indicates the direction of one's destination

Direct leveling: the process of measuring differences in elevation directly

Displacement: in photogrammetry, a shift in the position of objects in photos or images due to the perspective of the camera

Dot grid: a grid of dots used to analyze the area of objects in photos; each dot represents a proportional area in the photo

Down woody debris: fallen dead wood in a forest

Draw: also called a gulley or valley; the area of lower elevation between two ridges or spurs

Electromagnetic spectrum: the range of all possible frequencies of electromagnetic radiation from low-frequency radio waves to very high-frequency gamma and cosmic rays

Entity-object correspondence: a term frequently used in geographic information systems to describe the relationship between features on the earth and how they are represented on maps

Equivalent scale: uses two units to describe the relationship between map distance and earth distance (e.g., 1 cm on the map = 2 km on the earth)

Fiducial mark: an object or marker placed in the field of view over a photo or image; used as a point of reference for measurements

Frequency: the rate at which something occurs

Geographic coordinate system: defines location using angular measurements on a sphere (e.g., latitude and longitude)

Geometric dilution of precision: occurs in Global Positioning Systems (GPS) when satellites are very close together, degrading positioning accuracy

Geoprocessing: used in geographic information systems to manipulate spatial data

Georeferencing: the process of assigning coordinates to an image in order to identify its spatial location

GIScience: Geographic Information Science; a discipline focused on the collection, analysis, and improvements of geographic data using theory and logic from cartography, surveying, remote sensing, geography, mathematics, and statistics

Greenwich Mean Time (GMT): the mean solar time on the Prime Meridian (0° longitude)

Grid north: north of a particular geographic area represented on a map

Ground cover: the percent of an area covered by herbaceous vegetation

Gulley: also called a valley or draw; the area of lower elevation between two ridges or spurs

Image interpretation: the process of identifying objects in aerial imagery and determining their significance

Importance value: an index to assess the dominance or prevalence of a species in a community; higher importance values equate to a stronger presence

Indirect leveling: involves the calculation of elevation from slope and horizontal distances

Interval attributes: attributes that have a defined distance between values, and where the 0-value is arbitrary (e.g., time of day, calendar date, latitude, temperature)

Isocenter: in photogrammetry, the point halfway between the principal point and the nadir on a photo or image

Isogonic chart: a map or chart used to determine declination in a specific geographic area

Land cover: the classification of the earth based on the physical or biological cover over the surface of the earth, including water, vegetation, bare soil, and/or components of the human-built environment (e.g., roads or structures)

Land use: the classification of the earth based on human activities such as agriculture, forestry, recreation, or building construction

Land use and land over (LULC): a term broadly used to classify land into different categories; describes land cover, land use, or both

Large scale map: a map with a relatively large ratio between map measurement and earth distance; the features on the map are large

Laser level: a device used in surveying that projects a laser beam horizontally about a vertical axis

Latitude: the angular distance measured in degrees north or south of the equator

Legal description: identifies the location and boundaries of a parcel of land

Leveling: the process of measuring differences in elevation

Location: refers to a particular place or position

Longitude: the angular distance measured in degrees east or west of the Prime Meridian

Magnetic needle: the needle on a compass pointing toward magnetic north

Magnetic north: a location in the northern polar region that has a strong electromagnetic charge

Map: a model intended to depict specific phenomena within a geographic area

Map scaler: also called a planimeter; a tool used to measure the horizontal distances, perimeters, or areas of irregular shapes on a map

Mean: the central tendency in a distribution; calculated by summing a set of observations and dividing by the total number of observations

Median: the middle observation in a sample that has been ranked by magnitude

Midstory cover: the percent of the ground covered by the canopy of shrubs and small trees in a forest

Mode: the most frequently occurring value in a sample

Monument: in surveying, a physical object secured in the ground to mark survey points and lines

Nadir: in photogrammetry, the point on a photo or image directly beneath the camera

Navigation: a skill that involves the determination of position and direction

Nominal attribute: the descriptive information assigned to an entity (e.g., name, identification number, color, species)

Oblique photo: in aerial photography, when the camera lens is at an angle in relation to the ground

Observation: the measured value or characteristic of one sample unit

Ordinal attribute: implies ranking or order (e.g., suitability on a scale of 0–1; high-medium-low values)

Orienteering: a form of navigation that requires skill in the use of maps and compasses

Orienting arrow: an arrow painted on the degree dial of an orienteering compass; used to determine the direction of travel

Orienting lines: parallel lines indicating north–south direction on a map

Pacing: a method of measuring distance by using one's stride

Passive remote sensing: the acquisition of information by using the sun's light energy (e.g., photos and imagery are acquired using passive remote sensing)

Pattern: the arrangement or placement of objects in an area

Peak: also called a summit; the highest point in elevation

Photogrammetry: the science of making scale-accurate measurements from photos or images for the primary purpose of mapmaking

Photo interpretation: the process of identifying objects in aerial photos and determining their significance

Planimeter: also called a map scaler; a tool used to measure the horizontal distances, perimeters, or areas of irregular shapes on a map

Planimetric map: a map that only illustrates the horizontal position of features with no reference to contours, elevation, or other topographic relief

Plat map: a map that shows the ownership divisions of a piece of land

Point of reference: a stationary object that aids in determining position and direction

Principal meridian: in the US Public Land Survey System, the north–south line used for survey reference; divides townships between east and west

Principal point: in photogrammetry, the point where a perpendicular line projected through the camera lens intersects the photo or image; on a vertical photo or image, it is the geometric center

Public Land Survey System: a rectangular survey method to describe the location of land parcels in the majority of the United States

Random: without bias

Range: in the US Public Land Survey System, townships east or west of a principal meridian

Range: in statistics, the magnitude between the highest and lowest observations

Raster: a data model that represents entities as pixels or grid cells

Ratio attribute: values where the magnitude of difference is meaningful and can be added, subtracted, multiplied, and divided; cannot be negative (e.g., length, area, height, diameter, count)

Reflected energy: energy that bounces off objects

Refraction: the bending of wavelengths due to a change in velocity as it masses through different media

Relative: a metric considered in relation to, or in proportion to, something

Remote sensing: the acquisition of information from a distance

Representative fraction: the ratio of distance on a map to the same distance on the ground (e.g., 1:5000 means that 1 unit—such as inches or centimeters—on the map is equivalent to 5000 of the same units on the earth's surface); represents scale

Resolution: the amount of detail in a grid; the size of the pixels or cells within a grid (i.e., larger grid cells have lower resolution)

Ridgeline: also called a spur; occurs when land descends gradually from a higher elevation to a lower elevation

Ruler: a tool for making linear measurements

Sample: a portion of individuals from a statistical population that are measured for a study

Sample distribution: shows the frequency of the occurrence of observations within a given sample

Sampling unit: the individual from which a measurement is taken

Scale: the ratio between map distance and earth distance

Section: in the US Public Land Survey System, a parcel of land that is 1 square mile

Sextant: an instrument used for measuring angular distances between objects

Shape: the form of an object

Sine: the trigonometric function of an angle in a right triangle equal to the length of the side opposite the angle divided by the length of the hypotenuse

Site: an area of land and its associated features

Small scale map: a map with a very small ratio between map measurement and earth distance; the features on the map are small

Snag: a standing, dead tree

Spatial data models: used in geographic information systems to define the type of object used to represent entities; may be a vector or raster

Spatial thinking: a thought process that is used to solve geographic problems by interpreting the meaning of an entity's attributes, location, shape, and orientation on the earth

Spectral reflectance: the number of different wavelengths in portions of the electromagnetic spectrum (e.g., visible light, near infrared) reflected by objects

Spur: also called a ridge; occurs when land descends gradually from a higher elevation to a lower elevation

Standard deviation: measures the amount of dispersion that a sample dataset has from the mean of that dataset

Standard error: quantifies how precisely the true population mean has been estimated

Statistical population: the population that is the subject of investigation

Statistics: ways of summarizing and describing quantitative data in order to draw inferences on the data and understand whether the results are characteristic of a population or are due to random chance

Stereoscopy: a process for creating three-dimensional depth by viewing the same object in two photos from different vantage points

Summit: also called a peak; the highest point in elevation

Surface map: a two-dimensional representation of a three-dimensional surface (e.g., contour map)

Surveying: the technique and science of accurately determining the position of points and the distances and angles between them

Survey rod: also called a leveling rod; a graduated rod used to measure the vertical distance between a point on the ground and the surveyor's line of sight

Tangent: the trigonometric function of an angle in a right triangle equal to the length of the side opposite the angle divided by the length of the side adjacent to the angle

Target: a device used in surveying that is affixed to a survey rod and signals when the laser from a laser level is detected

Texture: the coarseness or smoothness of an object or area

Theodolites: a survey instrument used for measuring angles

Tiers: in the US Public Land Survey System, townships north or south of a base line

Tone: in photogrammetry, refers to the variations in the reflectivity of different objects

Topographic depression: an area with higher elevation surrounding it in all directions

Topographic map: a map that represents the hills, valleys, and other features of the earth's terrain

Topographic profile: a diagram of the cross-section of a topographic survey area

Topographic survey: contains information depicting the three-dimensional position of land features and the distances and angles between them; used for creating topographic maps

Topography: the description of the shape and elevation of features on the earth's surface

Total station: a surveying device that contains an electronic theodolite and electromagnetic distance-measuring instrument

Township: land parcels that are 6 square miles, or 36 sections

Training samples: in reference to land use land cover classifications, areas delineated within different land cover classes on an image for the purpose of assessing the range of spectral reflectance values associated with a specific class

Transmittance: the fraction of energy that is passed through something

Triangulation: the process of determining the location of a point by measuring the angles between it and the location of at least 3 known reference points

True north: the location where lines of longitude converge

Valley: also called a gulley or draw; the area of lower elevation between two ridges or spurs

Variability: the extent to which data points differ from each other

Variables: attributes that differ between individuals in a population that are measured to infer characteristics of the population

Variance: the standard deviation squared

Vector: a spatial data model that represents features such as points, lines, or polygons

Vegetation composition: the type and relative amount of vegetation that occurs in the study area

Vegetation structure: the way in which vegetation is arranged

Vertical photo: an aerial photo taken when the camera is directly overhead and the lens is perpendicular to the ground

Watershed: the area of land where all water under it, or draining off of it, goes into the same place

9 781516 565184

Printed by Libri Plureos GmbH in Hamburg,
Germany